PRAISE FOR *IN HOT WATER*

In Hot Water by Sridhar Deivasigamani is a fascinating narrative of innovation, challenging conventional norms and embracing change with resolute determination. Through a fusion of expertise, curiosity, and ingenuity, the author creates state-of-the-art heating solutions that not only leverage today's cutting-edge technologies but also foresee the evolving demands of tomorrow. This book serves as a beacon of inspiration, showcasing the transformative potential of innovation in preserving our natural resources for generations to come. I can't wait to see the next wave of innovations from the author and Intellihot.

—MICHAEL LUZ
Retd. CEO and President, Viessmann USA

An inspiring journey of curiosity and innovation that forever changed the way we think about the use of hot water in building environments. This book details Sri's journey and relentless pursuit in creating one of the most innovative and successful businesses in our industry. The book serves as a source of inspiration, offering valuable insights and guidance for other creative minds seeking new sustainable solutions and their own path to success.

— CRAIG WEHR
Former President and COO of Zurn Elkay

In this engaging narrative, Sri Deivasigamani unveils his remarkable story as a visionary entrepreneur hailing from the heart of India, whose unwavering commitment to innovation and sustainability is reshaping the water heating industry. His book chronicles the transformative journey of a true pioneer, whose entrepreneurial spirit and dedication to problem solving and environmental stewardship is leaving an indelible mark on both business and society. Sri recognized the pressing need for innovation in the water heating sector, which has not seen much change for a century. His book tells the tale of how he navigated complex market dynamics and a resistance to change to build a groundbreaking company that is setting new standards for energy efficiency, corporate responsibility, and environmental consciousness.

— ALEXIS ABRAMSON, PhD
Professor and Dean of Engineering, Dartmouth College;
Former Chief Scientist, Building Technologies Office,
Department of Energy

MIT prides itself on producing engineers who strive for solutions to global problems, all the while keeping in mind societal upliftment, empathy and care for others and sustainability. Sridhar's *In Hot Water* is yet another example of a young engineer from MIT fulfilling his duty towards mother earth.

— COMMANDER DR. ANIL RANA
Director, Manipal Institute of Technology

IN HOT WATER

SRIDHAR DEIVASIGAMANI

*a maverick's
cleantech journey*

IN HOT WATER

Forbes | Books

Published by Forbes Books, Charleston, South Carolina.
An imprint of Advantage Media Group.

Forbes Books is a registered trademark, and the Forbes Books colophon is a trademark of Forbes Media, LLC.

Printed in the United States of America.

10 9 8 7 6 5 4 3 2 1

ISBN: 979-8-88750-038-6 (Hardcover)
ISBN: 979-8-88750-039-3 (eBook)

Library of Congress Control Number: 2024905935

Cover design by Megan Elger.
Layout design by Analisa Smith.

This custom publication is intended to provide accurate information and the opinions of the author in regard to the subject matter covered. It is sold with the understanding that the publisher, Forbes Books, is not engaged in rendering legal, financial, or professional services of any kind. If legal advice or other expert assistance is required, the reader is advised to seek the services of a competent professional.

Since 1917, Forbes has remained steadfast in its mission to serve as the defining voice of entrepreneurial capitalism. Forbes Books, launched in 2016 through a partnership with Advantage Media, furthers that aim by helping business and thought leaders bring their stories, passion, and knowledge to the forefront in custom books. Opinions expressed by Forbes Books authors are their own. To be considered for publication, please visit **books.Forbes.com**.

This book is dedicated to those whose unwavering support and boundless inspiration have propelled our journey:

To Siva Akasam, my cofounder and CTO, whose patience, intellect, and unwavering enthusiasm have been the bedrock of our success. Your partnership is a gift I deeply value.

To Mike Gonzales, our chief engineer and design extraordinaire. Your multiple talents and unassuming, humble nature have transformed ideas into reality.

To Rod Harrison, our retired CFO at Intellihot, whose financial acumen and operational wisdom guided us through uncharted waters.

To Thangavelu, my grandfather, a humble but visionary farmer, who laid the foundation for our family's education. Your legacy lives on in every accomplishment for generations to follow.

To my father, Deivasigamani, whose hands-on approach, practicality, and hard work fueled my quest for engineering.

To my mother, Lakshmi, the source of my confidence and the embodiment of unwavering belief in my potential and inventions. You are my guiding star.

CONTENTS

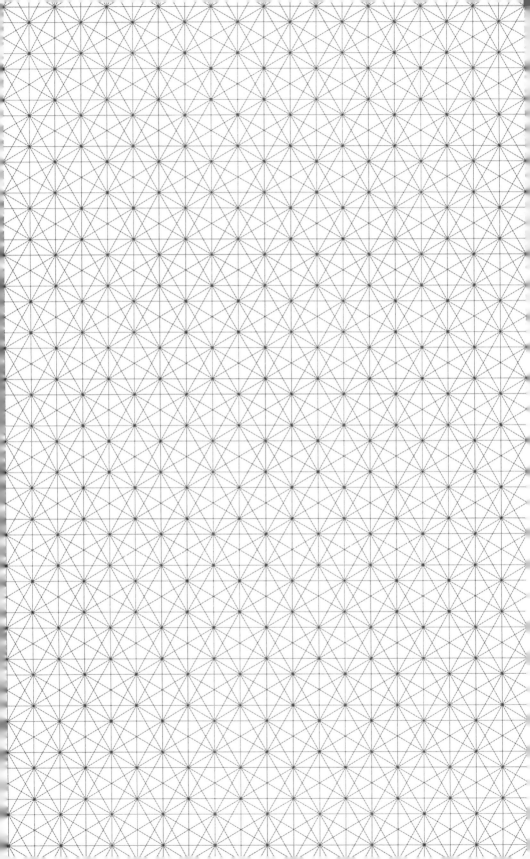

ACKNOWLEDGMENTS

This book stands on the shoulders of giants, and I owe my deepest gratitude to:

The teachers at RSK School, whose dedication laid the foundation of my knowledge.

The professors at Manipal Institute of Technology, particularly Professor K. M. Najundiah, for their tireless efforts to nurturing the minds of tomorrow's innovators.

The professors at Clemson University, notably Dr. Imtiaz Haque, for guiding me through challenging situations and helping me complete graduate school.

To my wife, Prithika, for suggesting that I write about my life experiences and work.

To my son, Kai, and daughter, Ihita, whose curiosities and laughter bring boundless joy, nonstop innovations of parenting skills, and purpose to my life.

To our faithful dogs, Whisky, Charlie, Fizzy, and Jewel, whose antics and love keep us grounded.

To the friends, investors, and supporters of Intellihot, whose belief in our mission has been the wind beneath our wings.

To my competitors, specifically AO Smith, Bradford White, Rheem, and Aerco. You have served our customers faithfully for decades and built great factories and enduring businesses. I gained knowledge and inspiration by observing your work. Thank you for being both aspirational and formidable contenders.

To the great community of Galesburg, whose spirit and support have been instrumental in our journey.

To Gretchen Carhartt, for her steadfast commitment to fostering clean and sustainable businesses.

To Cesar Suarez, for his tireless efforts in the economic development of Galesburg.

To Sal Garza, the former mayor, whose vision and leadership have left an indelible mark in the Galesburg community.

To Larry Hayward, friend and investor, and a chance encounter that led to creation of something great.

And finally, to all current and former employees of Intellihot, whose dedication and hard work have been the lifeblood of our success.

INTRODUCTION

We are all interconnected, extrinsically and intrinsically. We all live on the same planet, breathing the same thin sheet of air that surrounds our globe and drinking the same precious water. As per some estimates, billions of atoms in our body once belonged to someone else[1]—perhaps another human being, plant, algae, or dinosaur—the exact same atom passing through billions of years and generations without any change. There is no doubt we are all connected and affected by global events affecting our air and water; it doesn't matter if you live in a penthouse in New York or a rural village in India.

My awareness of this interconnectedness, of the importance of air and water, and of the egregiousness of its waste, all started as a kid spending time on my father's family farm in India, where there was no electricity or running water. But it was later in life that I discovered how I could really make an impact, when I found my true purpose, my true calling: reinventing technology in our built environment starting with water heating, an industry that had not seen any major innovations in a hundred years.

Finding that calling, the thing that would truly fulfill my soul, was a long and winding journey—a series of seemingly unconnected events that all came together into the innovative, groundbreaking clean industrial technology company that is Intellihot Inc. From growing up on my father's farm to my endless tinkering with batteries, lights, amplifiers, and radio transmitters, running experiments in my parents' house during childhood; from my love of motorcycles and intelligent machines to my degree in mechanical engineering from Clemson University; from my work on marine engines at Caterpillar to my collaboration with my friend and cofounder, Siva Akasam, whose skills and interests perfectly complement my own—each dot connected to the next to lead me to my true purpose.

The power of purpose, plus a healthy dose of self-confidence, enabled me to do the seemingly impossible: revolutionize an industry that had remained largely unchanged for the last century. There have been so many innovations in so many industries—the automotive industry, the energy industry, all of these industries that provide for our fundamental needs. And yet, the fundamental need of providing hot water in our buildings and homes—essential for human life and human sustenance—has not changed at all in one hundred years.

Because of this, creating an innovative new product in this industry, completely upending how things had always been done, was not a smooth and straightforward path. From experimenting in the basement to getting our product certified and bringing it to market, finding customers and investors, building a brand, and figuring out how to successfully scale up production—we were always pivoting, always reinventing, always finding new and unexpected ways to do things, always repeating our mantra: How hard can it be?

As I turned from inventor to entrepreneur to CEO, I learned countless lessons about everything it takes to run a business, bring an innovative new technology to market, and disrupt a consolidated industry with dominant players. But the most important lesson I learned was to always stay true to my passion, vision, and values.

I also discovered firsthand that "going green" or simply adopting a sustainable lifestyle doesn't always require spending more, changing behaviors, or being saddled with complicated tasks. It is entirely possible, and in fact conducive, to meeting bottom-line goals with positive environmental impacts. We have proven it at Intellihot, with products that cost less to own, cost less to operate, occupy less space, are more reliable and more efficient. They are safer and healthier for everyone involved. It's a value proposition that is economically compelling while also being vastly better for the environment.

Any time there is transformation of energy—whether to power our cars, heat our homes, or heat water—there is an impact on human health and safety. At that nexus of energy transformation, health, and safety lies the interconnectedness of air and water. This is where Intellihot is making a difference. Through my story, and that of Intellihot, I want to show you that you can make a difference, too—no matter who you are, where you live, or what you do for a living.

I want to reveal the power every single one of us has to make a difference, to take small actions that can truly change the world. I want to illustrate how we are all connected and how communication across industries, sectors, and nations could create change even more efficiently and effectively. And most importantly, I want to share the power of finding your purpose and the incredible things you can accomplish, against all odds and obstacles, when you are following the true calling of your soul.

All Great Ideas Start in the Basement

It was December 2005. I had just purchased my first home in Peoria, Illinois. I had never owned a home in the United States before—or, for that matter, anywhere in the world. It was three o'clock in the morning. And I was standing in my basement, ankle-deep in water.

I had just returned from visiting my parents in India. Flights to and from India are already very long, with multiple stops, and my hometown, Tiruchirappalli (or Trichy, for short), was anachronistically disconnected from the nearest international airport city of Chennai. It takes forty hours—a minimum of two days and two nights—to get there and the same amount of time to get back. Return flights from India back to the United States generally land very late at night. I landed in Chicago after midnight and drove over three hours to Peoria, arriving home around three o'clock in the morning. It had been a very, very long day of travel. To make matters worse, the eleven-and-a-half-hours' time difference was disorienting.

1

Usually when I get home from a long trip, the first thing I do is go down to the basement to make sure everything is alright. I picked this habit up from my friends who were homeowners, who advised me to always shut off the water during long absences. However, this being my very first extended trip as a homeowner, I had forgotten to do that. So, even though it was so late, and I was exhausted from the forty-hour international trip, I went down into the basement—only to discover that my entire basement was flooded.

I had no idea where the water had come from, but I had to quickly think of a way to deal with it. Just prior to my trip, I was shopping for a shop vacuum for my hobby projects. At Sam's Club, I saw a 5 hp vacuum categorized as "wet and dry." I thought, "Why would I need a wet vacuum?" but it was reasonably priced, so I purchased it. Little did I know how handy a wet vacuum would soon be.

I put the wet vacuum accessories together and started vacuuming up the water on my basement floor, pumping it twenty feet out of the house via a garden hose—which, thankfully, worked. As I was vacuuming, I discovered the source of the flood: my tank water heater.

I had never paid much attention to the water heater before. When I first bought the house, I went down to the basement to see what was there and noticed a big gray cylinder with two pipes, one going in and the other coming out. I thought, "This is simple; it must just turn on when I open the faucet. There must be some mechanism inside that triggers it on and heats that water up when I turn on the faucet." Until the heater leaked and flooded the basement—until I was vacuuming up water at three o'clock in the morning—I had never realized that the water heater not only held large quantities of water but also stayed on 24/7 in order to keep the water constantly hot.

After I vacuumed up all the water, I went straight to bed. I was too exhausted to do anything else, but my mind was restless thinking about that big, bulky tank. When I woke up in the morning, I was still bothered by the water heater. I was bothered by the fact that the water heater ran nonstop, using energy to keep this large quantity of water heated all the time—even when nobody was using it, even when there was nobody in the house.

A Colossal Waste

Waste has always bothered me. I have always had an awareness of the finiteness of energy resources. Even as a small child, I remember asking my father about what he put into his motorcycle to power it. "What is it called?" I asked. "Where does it come from?" He told me it was called petrol, which comes from petroleum. Then, I read that the word "petroleum" came from the Latin *petra oleum*, which translates to "rock oil." From there, I came to the logical conclusion that there must be a finite quantity of this substance.

"Aren't you worried that the petrol is going to run out one day?" I asked my father. "What are we going to do after that?"

4

I started reading about how the engine used its fuel. I discovered that only 20–25 percent of the energy actually goes to propelling the vehicle, while the rest is dissipated as heat. This means a colossal amount of energy—and fuel—is actually wasted. After this discovery, I would often tell my father that when he was coasting or going downhill, he should shift his bike into neutral, or even turn it off, to conserve petrol.

My father's motorcycle was just the beginning. I started noticing wasted energy everywhere. I became vigilant about turning off lights whenever possible, not running the tap when we didn't need to. This vigilance continued through my teenage years and into adulthood. Energy efficiency was always present in my mind. It still is! So, when my water heater leaked, I could not help but think about how much energy and water were being wasted.

When I woke up the next morning, I immediately began investigating whether there was a better option for heating water. I was shocked to learn that the method of heating water in our houses and buildings has remained almost exactly the same for over one hundred years, since the Industrial Revolution. You collect and store water in a large container; the size depends on the type of building. For a single-family home, the container is generally around 60–80 gallons. For a commercial building, it could be anywhere from 250 to 5,000 gallons. To heat the water in this tank, you either light a fire underneath it or install an electric element in it, and you keep the container of water hot 24/7—which means using energy 24/7.

Simply maintaining the temperature of the water in the tank wastes energy. We keep the water in these containers very hot, which means they have a high rate of heat loss no matter how insulated the container is. Consider a cup of coffee: the basic laws of physics dictate that the coffee will always get cold. In fact, the hotter the cup of coffee, the faster it will cool down—lukewarm coffee cools down at a slower rate. A very hot cup of coffee will cool down until it reaches a lukewarm temperature, and from there, it will take a longer time to become cold.[2] Maintaining a high temperature in the water heater tank requires a huge amount of energy simply to make up for that heat loss.

I also discovered that conventional gas water heaters, the most common type of water heater in residential homes, operate at only about 50–60 percent efficiency deliberately. The remaining energy is used for buoyancy of the flue gas, which enables it to be exhausted outside without needing a blower. Even Energy Star models have only 60–70 percent efficiency,[3] and the federal minimum standard for efficiency is only 59 percent.

I was outraged. It seemed so wrong. "My God," I thought. "Everybody has a bulky tank in their basement holding water. Everybody's heating element is on all the time to keep the water hot, sending greenhouse gases into the atmosphere. All day, every day. Now, multiply that by every household and building in the country. Hundreds of millions of fires just sitting and burning. What a colossal waste of energy!"

Separately, I started to notice how long it takes for the water to warm up when you turn on your tap. As you wait for the water to get hot, a stream of good, clean water is going down the drain, completely wasted. I would sometimes pull the stopper and catch the water in the sink in the hope it would hold my suds or give up its energy to heat up my house, but I knew this wasn't an effective way to conserve water or energy.

When I timed the faucets in my house, it took about sixty-five to seventy seconds for the water to get hot. I called my friend Siva. "How long does your water take to heat up?" I asked.

"I think it's around forty-five seconds," he said, "but I usually just let it run while I go grab a towel and get ready to take a shower."

"Human beings are so clever," I thought. "We fool ourselves into doing other things, while all this waste is happening right under our noses." People are conditioned for incompetence and poor performance. We do other things while we wait for the water to heat up, accepting that all of this water has to simply go down the drain, unused. We also accept that we have to wait. We don't know any better.

I called a friend in Boston and another in California. One of them lived in a condo building, and his water took almost ninety seconds to warm up. The other had a similar wait time. I called a few more friends in different places around the country and did a quick average of the wait times. If I added all the times together, at an estimated two showers a day, it came to about six months over the course of the average lifetime of a male in the United States. Six months of waiting for water to heat up. Six months of good, clean water going down the drain. Mind-boggling waste!

If the water in the container is kept hot, why does it take so long for hot water to come out of the tap? Because while the water in the container may be hot, the water sitting in the pipes between the container and the faucet cools down between uses. When you turn the faucet on, the system has to push all of that cold water out of the pipes before the hot water can come out of the faucet.

Most homes in the United States have their water heater in the corner or the center of their basement, while the bathrooms are scattered around different floors and corners of the house. A first-floor bathroom right above the heater will get hot water quickly, because the water doesn't have to travel far. However, it will take a bathroom on the second floor on the opposite side of the house a much longer time to get hot water.

I also realized that keeping water in large containers is, in fact, unhealthy. A bacterium called *Legionella* thrives in lukewarm temperatures, typically 90–110 degrees Fahrenheit. While a tank of water might be kept very hot, there are zones where the temperature is cooler. That, coupled with very low and nearly stagnant water velocity, creates a very conducive environment for bacteria growth.

I had noticed that whenever I traveled, if I drank hotel water, I would get a sore throat. Anecdotally, people had told me, "It's OK to drink cold water straight from the tap, but not hot water." But I never knew why. Now, I suspect it's because of the bacteria in hot water coming from the plumbing system.

On top of all of this, there was the aesthetic aspect: these water heaters are big, bulky, and ugly looking. Sometimes, you can also find dribs of paint left on the unit!

All of these issues stem from the system we invented back in the late 1700s and early 1800s. Since then, we have continued heating our water more or less the same way. I knew there had to be a better option.

A Smart Solution

Until my basement flooded, I did not realize just how much energy is wasted by our homes overall. On a fundamental level, from when we lived in caves in prehistoric times, human beings need three things to sustain: food, clothing, and shelter. We do everything we can to keep that shelter comfortable and safe. Today, in our shelters, we need energy for electrical power to heat and cool the shelter, and to heat water. Those are the three basic energy needs for our shelters—and in all of those areas, there is tremendous waste.

The furnaces in our homes create waste. The way we produce and transport electricity creates waste. When electricity is produced far away, you are only getting 30–40 percent of the original energy produced by the time it reaches your home. And the way we heat our water creates additional waste.

When I looked into how we heat and cool homes, it became apparent that these systems should be integrated. When a house is cooled, the heat is absorbed from the house and thrown outside. What if, instead, you could put that heat into the water and get hot water? I realized that, from the get-go, my idea was not just about a water heater but an entire integrated system. I wanted to build something that brought hot water, heated and cooled space, and produced electric power, all combined into one symbiotic device. But I knew, to get there, I had to start small. So, I consciously made the decision that the first problem I would focus on solving was water heating and, even more specifically, commercial water heating.

I could see that reimagining a centuries-old system was a challenging problem—made more challenging by the fact that I had no background in water heating, or the plumbing industry, or any building trades industry. At the time, I was experimenting with what I wanted to do in life. While working a full-time job at Caterpillar, I was also working on developing a smart motorcycle and selling efficient LED lightbulbs—a technology I had become a strong believer in twenty years ago, after realizing how much more energy-efficient and long-lasting LEDs are over incandescent lightbulbs. These diverse projects and interests combined to put me in a very unique position from which to consider this problem.

My work with, and deep interest in, intelligent machines led me to a seemingly obvious solution: an intelligent water heater. "Why can't we have a water heater that is intelligent and heats on demand?"

I thought. "A water heater that watches your behavior and can predict when you are going to be home and when you are going to need hot water? Imagine how much more efficient that would be, how much less water and energy would be wasted!"

I did some research and could not find anyone else working on—or even thinking about—a device that could not only heat and deliver water but also learn people's behavior and predict when they would need hot water. I was surprised because it seemed so logical to me. Human beings are predictable. Most people take a shower between six and eight in the morning or evening every weekday. An intelligent machine could learn this pattern and would know to heat water up at that time and transport it to the faucet. The rest of the day, the water would not be heated, saving a significant amount of energy.

I started investigating how to create a device that heats water on demand rather than storing it. At the time, I was working on a smart motorcycle, fitted with sensors that would anticipate velocities and braking and automatically shift for you, giving you the reflexes of a champion motorcyclist. I took the smart control system I was building for my motorcycle and used it to build a clever control system for a water heater. Within a few weeks, I had built a prototype of an intelligent water heater.

The name of the company was born from the idea of "intelligent heat." I was checking on the availability of domain names with various combinations of "intelli" and "heat." All were taken. Finally, I put in "Intellihot." The domain name was available, and that was that. Intellihot became the name of the company, and I knew I had found my mission.

The Intellihot system does not store hot water. Instead, when you open a faucet, the system detects it and heats the water on demand. When you're not using hot water, the system doesn't heat the water.

The heating unit shuts off. But the system goes further than that. The Intellihot system learns our behavior and predicts when you are going to use hot water. When it gets to the time that the system knows you are likely to use hot water, it heats the water and pushes it up to your faucet to wait for you. When you open the faucet, you get instant hot water—no time spent wasting water while waiting for it to heat up. It's not complicated, but it works like magic.

But coming up with this magic didn't happen overnight. It took my whole life—many, many disparate, seemingly unrelated pieces coming together into a perfect, magical storm. As Steve Jobs said, "You can't connect the dots looking forward; you can only connect them looking backward. So, you have to trust that the dots will somehow connect in your future." For me, those dots started all the way back in a township called Kailasapuram, near the city of Trichy, when I was just a small child.

CHAPTER TWO

Origins: Town and Country

Kailasapuram, where I grew up, was a very small, industrialized township. Our whole community was nothing but engineers and doctors. But just a day's journey away was my family's farmland near the Bay of Bengal, which could not have been more different.

My father was a mechanical engineer who worked for Bharat Heavy Electricals Limited (BHEL), a large energy company, similar to General Electric in the United States. BHEL was setting up many of India's thermal and nuclear power plants. While my father was an engineer, one generation prior—my family were all farmers, most of whom didn't make it past fifth or sixth grade. In fact, there was no school in the village where the farm was located. My father went to school 30 km away.

My grandfather, my father's father, barely finished fourth or fifth grade himself, but it was very important to him that all of his children—three daughters and twelve sons—were educated. Almost all of the children were sent to a school that was located outside the village. The first was an arts major who was then drafted into the Rapid Action Force. The second was an attorney. The third and fourth were engineers. The fifth was a chartered accountant, and so on. They all went from rural farm life with very little education to being highly educated professionals.

However, lack of education didn't stop my grandfather from being an innovator. He was always poring over magazines and reading articles. As the story goes, my grandfather imported the first mechanized tractors in South India from a company in Belarus called Minsk Tractors, which still exists today. When the tractors arrived, my grandfather brought them out into the field—and they promptly sank in the South Indian clay. The tractors were too heavy for our rice paddy fields! So, he experimented and built several large, grated steel wheels, distributed the weight, and somehow made it work.

One day the tractor had trouble starting. My grandfather called his middle-school sons, thinking, "They're all in school; they're smart now!"

"Ah yes," my father said, "I know how this works!" But in truth, he didn't know anything about internal combustion engines. His knowledge was related to steam engines. With that knowledge, he proposed: "If we collect all this exhaust and put it in a drum, we can put it back into the tank and recycle the fuel." My grandfather, who also didn't know how internal combustion engines worked, thought this was an excellent plan. So, they collected all the exhaust gas, put it in a drum, and tried to recycle it—which, of course, did not work. My grandfather quickly chased all the schooled kids away! This is the spirit that runs in my family, which has brought me to where I am today: we are always trying new things, new ideas and solutions, even if we have no direct knowledge of the matter at hand.

My father's family farm is also where an awareness of energy efficiency was instilled in my mind. While we lived in the township, my father would frequently take us to the farm where he grew up. Because of this, I grew up with two backgrounds: one very rural—swimming in rivers and ponds, walking through fields, fixing tractors, with no electricity and not even a real bathroom; and one with a very westernized, industrialized lifestyle with plumbing and electricity. In town, there were motorcycles and cars. On the farm, I would travel with my grandmother from our village to the next village in an oxen-driven bullock cart. That was our main form of transportation!

On the farm, with no electricity, we did everything by hand. I picked up many lessons on how to do more with less. I could even make grease from hay! My Chittapa (uncle), in particular, was very scrupulous about how he used energy. He would spend all day working in the fields on motorized pumps for field irrigation; standing water being essential for paddy fields. As evening approached, my uncle would say, "Don't turn on the torch (battery flashlight) until we absolutely need it! We can still see without the light for another few hours, and we don't want to waste energy. If we turn on the light early, our eyes will get used to it and we will need keep running the flashlight longer." We would let our eyes acclimatize to the darker and darker evening, until we absolutely couldn't see, and only then was the torch turned on.

I never took electricity, water, or energy for granted—which highlighted for me how much energy waste existed in more developed societies. If I had lived only in the more modern township, without a rural farm life, I likely would not have noticed the contrast as sharply, either in my childhood or as I grew older.

This awareness extended throughout my childhood and young adulthood in India—sometimes leading me to some unusual behavior!

For example, I knew that fluorescent lightbulbs provide more light while consuming much less power than incandescent bulbs. So, when I went to college and saw incandescent bulbs in my dorm room, I wanted to replace them with fluorescent bulbs. However, fluorescent lightbulbs—or tube lights, as we called them—were expensive, and I was a student without a lot of expendable cash. So, I developed a simple high-voltage inductor circuit that could take a blown tube light and make it usable again.

Tube lights have two filaments, one at either end. They become unusable when one filament is blown, usually turning that end black. At that point, the bulbs are usually discarded. I would go around the campus and collect the discarded tube lights and make them light up with my special circuit. I was very happy because I was able to change out the 100-watt incandescent lamp for a 40-watt fluorescent lamp, with which I got more light while consuming less power. The best part was the tube light was free! This collection of blown and blackened tube lights was my prized possession. Each time I moved dorms, I would bring this big bundle of blown fluorescent bulbs—from my first-year dorm to my second-year dorm to my third-year dorm. Not the average accessory for a college student!

College was far from the beginning of my career as an inventor. In fact, I had been experimenting and tinkering since I was a child.

Childhood Experiments: Disco Lights, Perpetual Machines, and an Engine That Ran on Water

From a young age, I was very interested in planes and motorcycles and engines—anything that had a combination of engines or mechanicals and electronics. I was always taking things apart—flashlights, cassette players, any kind of toy I had—in order to get at the circuit boards or motors inside. There was never a working radio in the house! I would use the motors to build other small, battery-powered devices like fans, cranes, planes, or boats using shoeboxes, talcum powder containers, and the like.

One of my favorite quests was to build a perpetual machine, one that supplied unlimited electricity. When playing with DC motors, I found that if I took a battery and connected it to the terminals of a motor, it would spin. Then, I discovered that when I connected a lightbulb—the kind you get from a flashlight—to the motor terminals, I could make the bulb glow by spinning the motor.

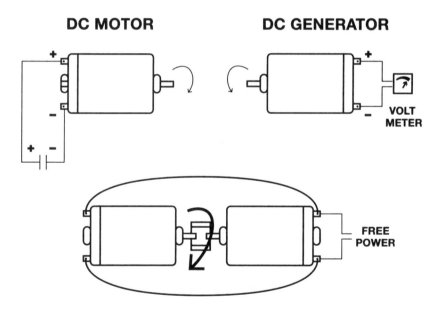

From this, I learned that a DC motor can become a DC generator. "What would happen if I connected two motors together on a common shaft and connected the outlets from one to the other?" I wondered. "It seems like it would send power from the generator to the motor, cause it to spin, and drive the generator, which would produce power, which is supplied to the motor, causing it to spin the generator, and the cycle would continue endlessly. To start it, I could give it a little spin, but it should then keep on running on its own. Also, I can draw a little power off the generator to power other devices, and then I'll have perpetual power!"

I would constantly draw diagrams of this perpetual machine. I even built a couple and tried hard to make them work, but I was never successful. The machine would work as a generator and as a motor, but I could never get it to work as a perpetual machine. In third or fourth grade, when I was attempting to build this device, I didn't understand the concept of energy conservation—that energy can neither be created nor be destroyed—or of efficiency losses. By the time the mechanical energy gets converted to electricity, there is slightly less energy than what went in. When that electricity becomes mechanical, it loses even more energy. There are always losses, which is why the machine could never keep running perpetually. I came to understand these concepts later, but as a little kid, I was very perplexed.

Also, by third or fourth grade, I was playing with high-voltage—230 V, 50 Hz—electricity. My parents were either oblivious to this or had a high degree of confidence in my abilities!

Of all my inventions, my favorite was building disco lights when I was in third or fourth grade. To make four lightbulbs light up in a repeatable sequence, I used a Ponds (a household brand) Talcum Powder cylinder. I scratched off the paint in a circular fashion, revealing the bare metal but leaving multiple sections intact. I then made four more of these circular sets and one for power, in which was a fully bare circle. I hammered a series of nails on a board, put the cylinder in between, and connected supply wires to the nails and to the four bulbs. Then, I rotated the cylinder using a wooden tree branch, which acted as an insulator. Rotating the drum brought a specific pattern of bare metal patches into contact with the nail, lighting up that specific bulb. Essentially, the metal cycling received electric power, and then it transmitted the power to each of the four bulbs as the drum rotated, and as each of the supply nails came in contact with a bare section on the drum. I could make the lights flash intermittently based on the pattern I carved on the can and make it go faster or slower by changing the speed of rotation. I even took the contraption to my school to make the class fancy-dress competition more interesting with lights.

Of course, the whole thing was dangerous, but I was very careful with electricity. I recall while I was making this in our garage, my friend Vasudevan picked up a pair of heavy metal tongs and was about to hammer a nail while it had power. Luckily, I had the presence of mind and managed to stop him in the nick of time. To this day, I consider this one of my greatest inventions—and one of those few moments where my gray cells actually fired correctly!

Like my father and grandfather with the diesel tractor engine, I was always coming up with theories on how things worked, even if I knew nothing about the subject. My town was located near a major railway junction, which meant there was a big railway station nearby. The railway community hosted a science fair at a local college, and my school was invited. I went to the fair with an electric boat I had built, complete with a propellor, a motor, and a battery.

At this college fair, as I was showing off my little boats and motors, I got into a conversation with a visitor about how diesel locomotives work. I knew they had diesel engines that powered generators which produced electric power. This electric power was sent to electric motors, and it was the motors that moved the engines and carriages. I also knew, from my little experiments, that motors could be turned into generators that could generate electric power. So, I postulated a theory on how locomotives must work: when they had to move forward, they would connect the motors to the diesel-powered generators, and when they had to brake, they would just disconnect the motor from the generator and connect it to a bank of batteries. In my mind, this would happen via a double-pole, double-throw switch.

Then, the motor would become a generator and generate power, which could recharge the battery. It would also provide the retardation to brake the locomotive. This is essentially the regenerative concept used in hybrid cars like the Toyota Prius today—and I had postulated this as a kid! It was crystal clear in my mind that this must be how locomotives operate: when they have to move forward, they connect the batteries. When they have to brake, they change the polarity and the braking motion charges the battery.

I finished explaining my whole theory, and the person I was talking to said, "You know, I actually work in the locomotive repair shop, and I have to tell you: locomotives don't work that way." I can't tell you how disappointed I was. I thought, "Why don't they? That's how they should work!"

Throughout my youth, I never abandoned my inclination to follow crazy theories and ideas, wherever they led. I built mini rockets using the steel jotter from ballpoint pens. For one, they already looked like rockets, and besides, they were made of metal and had a plastic bottom with a small hole, which seemed perfect for a thrust nozzle. Of course, when I filled this thing with petrol, it would just fizzle, lay on the ground, burn, and melt the plastic end!

As I got older, around eleventh grade, I took to motorcycle riding. I did not get a license as I was not old enough and also because it was a cumbersome process filled with red tape. Nevertheless, I liked to explore the town on my motorcycle, going down roads I'd never been on before. One day, I went down a dirt road, following it until it opened out into lush farmland. I loved seeing all that green, after spending most of my time in the densely packed urban environment of our town. I started going there more often on my motorcycle, just to enjoy the farmland views—until one day, I went a little too far and ran out of gas miles away from the nearest gas station.

As I sat wondering what to do, it occurred to me that I did, in fact, have gas in my tank—but the level was just below the fuel petcock inlet, which was raised. How could I access that last bit of fuel? I stood on the road, looking out over the rolling fields of rice paddies, watching the large borewells pump water into the paddy fields—and it hit me: water is heavier than fuel. If I put a little bit of water into the tank, it would sink to the bottom and lift the fuel up to the level of the fuel petcock.

I scanned the fields and spotted a farmer standing in the field. I walked over to him and asked, "Could I have a little bit of water? I need to put it in my bike." The farmer gave me a strange look but let me take some water. I poured it into my tank as the farmer looked on, clearly thinking, "This guy is bonkers." Then I pumped the engine—and it fired.

"Hey!" the farmer yelled. "Have they figured out how to run a motorcycle on water now?"

I don't remember what I responded; I was so happy, I took right off, leaving him perplexed. As it happened, a few years later when I was in college, the same thing happened to a friend and me when we were out on our motorcycles in the mountains—and I amazed him by using the same trick I had discovered in eleventh grade!

Growing up, whenever I was put in a corner, I tried to get myself out with whatever I had within my eyesight—a useful skill I've kept with me throughout my life.

Free-Range Parenting

All of these experiments and experiences were only possible thanks to my parents. My family was not rich, but we were solidly middle class, and I was fortunate that my parents were able to provide me with food, clothing, and shelter. My parents also afforded me great freedom as a child. They did not overly involve themselves in what I was doing, instead practicing what I have come to call "free-range parenting." As a child, I already loved traveling—a love that has continued my whole life. By seventh grade, I was visiting all sorts of faraway places by myself. Most importantly, my parents allowed me to conduct all those crazy experiments.

Although their parenting was "free range," my parents were big influences in my life. My father used to travel for work, and when I was in first or second grade, he brought me a very simple balsa wood glider—the kind with a rubber band connected to the propellor you could wind up. This ignited my interest in airplanes; I wanted to learn everything about them. I remember asking my dad how lift works, and he sat down and explained Bernoulli's theorem of lift to me.

After this, I was always building model airplanes. My father's eldest brother, the firstborn son, was drafted to be a pilot with the British Airforce in World War II—one of millions of Indians drafted to serve in World War II who have long been unrecognized. During his service, when he was only twenty-one or twenty-two years old, he was shot down, and our family never saw him again. When, as a child, my affinity for airplanes, engines, and electronics emerged, my parents thought I could be a reincarnation of my father's older brother. They went as far as pasting the picture of a Hindu god over a photo of my uncle to hide his photograph on the wall. Perhaps they thought I would recognize myself—I don't know!

When I was a little older, my father was working on his motor-cycle. He asked me for some help setting the contact breaks or points on his motorcycle. I asked him what the points were, and he explained how they set the ignition timing for the engine. This set off my interest in how engines worked and in motorcycles—both of which would play a big role in my life and in the Intellihot story.

My mother was not an engineer or a scientist, but she was always my number-one fan. "My son is a scientist," she would say. "He's going to do something great." Even though she didn't understand what I was working on, she was always a big supporter—even when it put her in danger!

During my earliest experiment as a child, I remember sitting in the living room where my mother had just mopped. I had taken apart an electric iron because I wanted the copper wires, which were hard to find. The cord had a three-point plug, and in India, the voltage is 230 volts—higher than it is in the States. There were three strands of wire—one red, one black, and one green—each covered in rubber insulation and a little cotton sleeve. I had peeled off the insulation so I could have the three exposed strands of wire—essentially just a cord and a plug with bare ends—coming down to the ground.

I don't remember exactly what I was trying to accomplish with this experiment, but I remember telling my mother, "I want to try something, but I'm afraid to turn the outlet on. Can you turn it on for me?"

And my mother, who thought of her son as an "inventor and scientist," said, "OK." She used the wooden handle of her broom and flipped the switch—and the outlet went bang! There was a little black scorch mark around that particular outlet for many years. Looking back, I'm very grateful my mother didn't get electrocuted—because the floor was still wet from her mopping earlier.

I had blown a fuse. So, I went down to the mechanical room, which was away from the house, and changed the fuse. In those days, you couldn't find actual fuse wire. So, people would cut a piece of regular wire and use that as a fuse—which was, of course, not the right way to fix a fuse, but it's what we did! Lucky me, I had just landed a long cord and metal wires.

My mother's consistent and unequivocal support made a huge impact on me. She instilled in me a positive can-do attitude and a belief that I could do anything I set my mind to. She would often say to me, "Son, you are smart. You are intelligent. You are going to do something big for the world." This gave me immense self-confidence that allowed me to keep going, no matter what. Without that self-confidence, imparted to me by my mother, Intellihot would never have come into being.

School Days

While I was always learning out in the world and through my little experiments, school itself was another matter. My parents never explicitly told me, "You should study. You should get good grades." It was an unspoken expectation. But I would get bored because in school, they didn't teach what I was interested in learning.

In India, classroom sizes are large. I was in a class of sixty-five to seventy students. My school, RSK, was named after RS Krishnan, the first general manager of BHEL, the company my father worked for. It was a great place of learning with several dedicated teachers. My love for physics was, in part, due to a brilliant teacher, Mr. Rajagopalan.

From kindergarten to high school, two girls (Savita and Renuka) in my class always ranked first and second—and every single year, I ranked twenty-sixth in the class. This mid-level ranking never bothered me; I knew I understood the concepts, and I was just bored. In sixth or seventh grade, I decided to see if I could improve my rank. Nobody told me to do this; my parents were fine with my ranking, as long as I wasn't failing. I simply wanted to test how well I could do. Sure enough, when I really put the effort in, I shot all the way up to second rank—much to the shock of the two girls who always ranked first and second and to their worry, as I was only a mark or two away from being the first-rank holder! I did this just to see if I could—and to boldly do something without worrying about my image. Once I'd proven to myself that I could reach a high rank if I wanted, I felt no need to keep trying, and the next year I promptly went back to twenty-sixth.

My nonchalant attitude toward my studies continued through the rest of high school. In twelfth grade, we were required to take board exams, which are a big deal in India, as they can determine college admission and thus, the rest of your life. It was the day before my board exam in math. Usually, students hunker down and study hard, with their parents feeding them food and coffee and ensuring they stay on task.

But me? I was working on my true passion: my motorcycle. My mom peeked out at me from the balcony, down to where I was working in the portico, my hands fully greased up. "Do you want to be a mechanic?" she yelled down at me. "Or are you going to go study and do something else with your life?" This was the only time my mother ever explicitly said something about my education—and I was shocked. But I am glad that I listened to her. I temporarily turned my attention to my studies, did well on my exams, and was ultimately accepted to Manipal Institute of Technology, where I pursued a degree in mechanical engineering.

Manipal Institute of Technology—or MIT, as it was often referred to—is probably one of the best educational colleges in India. It has a diverse array of students from every corner of India along with numerous dedicated staff and teachers. From Professor K. Kamalaksha to K. M. Nanjundaiah to B. G. S. Prabhu to Alladi Prabhakar and others, all had a tremendous impact on our future lives.

At MIT, I continued to follow where my interests and curiosities led me—whether that was related to my schoolwork or not! One of the interesting decisions I made for myself while in college was that in those four years, I would pursue what I was interested in and what I was passionate about. I wasn't going to do what people typically did in college—dating, for instance. I thought that would be a distraction. This turned out to be one of the best decisions I made, because it allowed me to focus on the things I loved.

Two of those things were the same passions I'd had in my high school days: motorcycles and traveling. In college, I had a Yamaha RD350. It was a two-stroke powerhouse but had terrible mileage—only about 12 or 13 km/L. From my hostel (dorm), the nearest gas station was 10 km away, meaning I needed a liter to get to the gas station and one to get back, which left me only 3 L in reserve. This strongly motivated me to try to make my bike more efficient—which became an ongoing quest throughout my undergraduate years. I built many gadgets—high-energy electronic spark ignition, rejetted carburetor, improved timing, running twins on one carburetor with a common manifold, engine timing adjustments, and more—to try to improve the efficiency of my bike. Unfortunately, for all my love and care, my motorcycle was better but still very inefficient.

My love for motorcycles led me to keep my bike in my dorm room, right next to my bed. Yes, I slept with my motorcycle next to me! Since the gas station was so far away, and my bike was so inefficient, I also kept a can of gasoline in the dorm room. One day, my roommate, Ramnath, finally asked, "Hey, what is the stuff you have in the red can? It smells like gas. Is it OK to have gas in here? Isn't it dangerous?"

"Yes," I said. "It's very dangerous."

"How dangerous?" he asked.

"Watch this."

I took a cap full of gas, put it in the toilet, lit a match, tossed it in—and the whole thing burst into flames a couple feet high.

"Oh my god," my roommate said. "This thing can burn a building down! This is what you're keeping inside our dorm room?"

"I'm very safe with it!" I insisted. "It's all the way over in this corner!"

Perhaps against his better judgment, my roommate allowed me to keep the gasoline—for which I was very appreciative! We are still great friends today.

When I finished college, I did not start out with the idea that I should come to the States. Many of my friends were taking the Graduate Record Examination (GRE) and the Test of English as a Foreign Language (TOEFL) and applying to graduate programs, but I was very anxious to be done with school. I felt like I had to go out and start something of my own. I knew I didn't want to go back to my hometown; I wanted to be out on my own for a while.

My college, like many colleges, held campus interviews in our final year, at which you could interview with different companies and, ideally, get hired. However, there was a restriction: you could only interview with a certain number of companies, and once you were offered a job, you couldn't interview at any more companies.

I planned to use my interview with the first company (Incorporated Engineers Ltd., Baroda) as a mock interview, a sort of practice for the next interviews with the companies I was more interested in. But things didn't go as planned: that first company offered me a job, and I wasn't allowed to interview with any other companies.

This company was located far away in the northern part of India. "At least it will be a change of scenery," I thought. So, I took the job, and I moved somewhere I didn't know anybody. The company was very nice; they gave me an apartment in a building with all the other new employees. But as I was walking toward the head office on my very first day, I had a sudden, surreal feeling that all my years of education lay behind me and I was now stepping into the real world, literally on the road as I walked—and into the rest of my life, laid out ahead of me. My education, passions, and curiosities were in the past.

"Is this it?" I asked myself. "Is this all there is? Am I done?"

That very evening, after finishing my first day of work, I concluded I wasn't done learning. I went and obtained the application for the GRE and TOEFL. I continued at my job for about four months, until I took my GRE, and then left the job and returned to my hometown.

While in my hometown, I was able to foster another part of myself: my entrepreneurial spirit.

Rock-Cutting Burners

I've always had an entrepreneurial spirit—another characteristic my father and I have in common. When I came back home from Baroda, my father told me about a unique problem his friends in the granite mining industry were facing. Most of the granite countertops you see in the United States come from India, as well as Italy and South America. The granite is mined using a rock-cutting jet burner, which is used to spall the rock out of the rock face by a process called jet channeling. This jet burner is a very specialized device, about eighteen inches long and six inches in diameter, mounted on a steel pole about six to eight feet long, powered with kerosene and high-pressure compressed air. When used, it sounds like a jet engine. Jet channeling greatly sped up granite mining—you could cut rock in two days instead of the five days it took with a conventional quarry drill.

This specialized jet burner was causing my father's friends grief. The burner my father's friends were using, which they imported from Japan, didn't last very long and was very expensive. As my father described this problem to me, I thought to myself, "Hm, I could take this problem and turn it into a business." I had entrepreneurial spirit, and my dad was a good metallurgist. So, with his help, I designed a high-temperature burner head made primarily out of Stellite (a high-temperature and abrasion-resistant steel alloy) and a better jet burner. Thus, I started making India's first rock-cutting burners and selling it to a few people on the side.

In order to turn this into an actual business, I had to establish it as a small-scale industry (SSI). This turned out to be quite a cumbersome process—at least it was back then; it may be much better now. In order to complete the process, you have to register your company and open a bank account to collect payments. In order to do this, you have to go to the village tahsildar—essentially the village administrator.

I was young, fresh out of college, and excited about this venture. So, every day, I would go to the tahsildar's office and wait. And every day, the secretary would say, "The tahsildar cannot see you today. Come back tomorrow."

At first, I didn't understand why I was not allowed to see the tahsildar. Then, after a couple rounds of being told, "Come back tomorrow," I realized that the secretary was expecting me to grease his hands in order to get an appointment.

I was outraged. "I'm not doing this," I said. "He's supposed to do his job as a government official. I am not going to bribe him to do the job he is already paid to do." So, for the next month, I showed up every single day. It wore me out, but in the end, I finally saw the tahsildar and acquired what I needed to establish my business as an SSI.

However, I felt disappointed, even disenfranchised. I was trying to do something on my own, something that would benefit the country, but I was being impeded by a corrupt system.

I felt like I was at a crossroads. My college education had come to an end, and I felt as though I had a big journey ahead of me—but I wasn't sure if rock-cutting burners was the path I was meant to follow. I also didn't like dealing with corrupt bureaucracy. But starting a business—that was energizing. A thought struck me: "I've heard the United States is the place where you can start a business easily. I'm going to go to the USA!"

And that was it. With a reinforced desire to continue my journey, I started working on applying to graduate school in the United States.

Grad School Application Adventure

Applying to schools is an elaborate process. You take the GRE, you take the TOEFL, you complete the application, you get admitted—and then you confront the financial questions: Will I get some financial assistance? Is it a teaching program, or do I have to pay? I come from a middle-class family; I couldn't afford full tuition. On top of all of that, I was determined to study what I was passionate about. It was a lot to balance!

I went to the US embassy in Chennai and obtained permission to use their library. I would flip through the Peterson's guide and look at which schools offered programs I was interested in. My love for motorcycles was still burning strong, as was my passion for airplanes. So, when I started applying to graduate schools, I pursued two interests: aeronautical engineering and automotive engineering. And I actually got admitted to programs in both! Then, I tried to pick the schools that were most aligned with my interests—and ones where I could land a scholarship.

I applied to eight schools and got admitted to six, including Syracuse, Clemson, Wichita State, and the University of Alabama. They were all in wildly different locations, but I had no preference on that. When you are in India, America feels like a concept, not a real place. I had no idea how distinct the different regions of the country are. I had no idea the extent to which the weather and landscape and culture vary. I had no idea whether I wanted to be in Alabama or New York—I didn't know the difference! So, I paid little attention to where the schools were located.

Originally, I had decided to pursue a master's in aeronautical engineering at Wichita State—known as the airline capital of the world because it is the home of Cessna and Learjet. But when I told my parents, my mom gave me a look that clearly said, without words, "You're going so far away. I'm worried about you. And I don't want you flying planes."

"Fine," I said. "I won't do aeronautical engineering. I'll just do regular mechanical engineering." So, instead of going to Wichita State to pursue a master's in aeronautical engineering, I accepted an offer to study mechanical engineering at Clemson.

But the application and decision process were just part of the adventure. In order to study in the United States, I also needed to show financial viability and get a student's visa, called F1.

Dhobi and the Visa Quest

Back then, trying to get a visa to come to the United States was a horrific process. It generally took five or six months, and in my home state, you had to stand in line for three nights and three days—no exaggeration—just to get an appointment. People would literally camp out outside the US embassy building day and night in a line.

There have been many individuals in my life who have helped me along, but I will never forget one person who was instrumental in this part of my journey. At the time, my father was running his own start-up enterprise, a high-tech fabrication shop where he manufactured the high-pressure valves that go in power plants. The valves required some complicated geometry and high-temperature Stellite welding.

At this time, there was a young dhobi (washerman) who came by the house to collect my father's clothes, launder them, and bring them back clean and ironed. With everything my mother did, taking care of the house and four children, it was helpful for her to outsource this bit of work in this stage of her life. My father took a great liking to this young dhobi, who was partially deaf and had never been to school. He invited him to apply for a job as an apprentice machinist at his factory. My father paired him with a very senior machinist, and together, they became the best machinist team at the factory. The dhobi was very grateful to my father for giving him this opportunity and setting him on this career path.

When it came time for me to go through the grueling process of obtaining my visa, the dhobi saw his chance to return the favor: he offered to take the nights standing in line, while I stood in line during the day. He stood in line for all three nights, allowing me to get my visa—and also get some sleep!

I am incredibly grateful to this man, for his help in the visa line and also for what he taught me: that when people are given opportunity, along with mentorship and training, they will rise to it irrespective of their education level or background. I've kept this in the back of my mind and try to put it in practice even now. There are people in my company who had never done the job they're doing now, until we put them into that position. For example, we have a woman who had never worked in supply chain and demand flow until we moved her there from another department—and she has been terrifically successful.

The process of waiting in line got me in the door to the visa appointment; the appointment itself was an additional obstacle course and a nerve-racking one.

It Takes a Village to Go Overseas

The dhobi was not the only person who helped me out in my journey toward graduate school. Once you have an appointment for your visa, there is another hurdle: demonstrating that you have enough funds to come to the United States during the visa interview. If you are rejected once, it is nearly impossible to apply again. I'd also heard there was one official at the embassy who almost always rejected people.

Often, people don't have sufficient funds—myself included. So, what do you do? You beg and borrow from friends and family and neighbors. You ask, "Can I borrow a thousand bucks?" you deposit that money into your bank account for one day, and you repeat until you have enough. I must have had at least ten to twelve people who lent some money. Once the deposits are done, you ask the bank for a letter stating your large bank balance. You take that letter and use it as proof that you have sufficient funds to study in the States. Then, you return all of the money you borrowed—plus interest for one or two days. Even that amounted to a sizable sum!

At the embassy on the big day, my nerves were running high. I didn't have a plan B; coming to the States was my only plan. What would I do if I didn't get my visa? "Well," I thought, "we'll cross that bridge if we come to it." Thankfully, my appointment was with a nice official—and I successfully obtained my visa. As I left the appointment, I was overwhelmed by a sense of how my whole life was about to change.

With this ordeal finally over, I walked out of the embassy to where my father was waiting and shared the good news. I was hoping for a celebratory lunch. Instead, he said, "You've got your visa. Now, let's go book your plane ticket." Once the visa interview is complete, if your application is approved, you have to return to the embassy in the evening to collect your visa and passport. There was a gap of three hours before I needed to return, so we walked straight from the embassy to a travel agent's office—a distance of a kilometer or two—and booked my flight. I reckon my father didn't want to risk not getting a plane ticket. With this, I picked up a lesson from my father: don't procrastinate or postpone. If something is crucial to your success, just get it done as soon as possible. We did have a celebratory lunch afterward.

My coming to the States was enabled not just by my parents but also by many people in our neighborhood and community who helped contribute to this effort.

My acceptance in hand, my visa obtained, my finances … well, I'd worry about that later. I was ready to head to the United States of America.

America Is a Whole New World

Preparing and traveling to the United States was a blur. Everything was new. Despite my fascination with airplanes, I had never flown before! I had often stood outside the airport to watch the planes take off and land, but I had never gone inside the airport. I'd never been through immigration. My first experience with all of this was lugging two big suitcases my family had packed full of things they thought I would need. They even sent me rice and spices—although I had little experience with cooking!

By the time I got on the plane and sat in my seat, I was so exhausted that I passed out immediately. The next thing I remember is getting ready to land in Frankfurt. From Frankfurt, I flew to Atlanta—landing right in the midst of the 1996 Olympics, which were hosted in Atlanta that year.

I remember sitting in the airport and seeing two or three other people, who were also students, looking through and trying to make sense of the currency. I joined them, and we went over to the coffee shop. The menu said coffee cost $1.50, so we immediately made a mental calculation of how much that equaled in rupees, whether we had enough to buy a cup, and if it was worth the money.

I was truly a stranger in a strange land. Every time I've moved in my life, I've gone someplace where I knew absolutely nobody. Even for my undergraduate degree, my first time ever being away from home, I moved two states away. It took two nights and two days to get to my school. I knew no one. They spoke a different language and had different customs and culture, so everything was different. When I had my first job in India, I went to the state of Gujarat, in north India, and it was the same story there. And when I moved to the United States, I didn't know anyone.

Thankfully, Clemson had an India Association that helped new students coming in from India. Their help included providing room and boarding for the first couple of days while we got settled. While I was looking for an apartment, I used a senior's bedroom as temporary housing. I remember looking at his air-conditioner and thinking, "I've never had an air-conditioner." That thing was cold and made a racket! With all the unfamiliar elements, I could hardly sleep at night.

My sleeplessness wasn't helped by the fact that mattresses are very different in the United States. The senior's bed had a box spring and a mattress, which I'd never seen before. I tried to sleep on the mattress but found it was much softer than what I was used to, and so I couldn't get comfortable. So, I pulled off the mattress and felt the box spring—having no idea what it was—and said, "Ah, this feels hard enough for me." I threw my sheet over it and, not knowing any better, slept on just the box spring!

As an immigrant, there are many little quirks as you adjust to a new place, a new culture, and a new lifestyle. I would look at large double doors and think, "How do I open the door?" because even the door handles looked so different. I went to my first day of classes, and one of my professors—Dr. Harry Law, who would later become my advisor—was wearing shorts. I was shocked—in India, you would never see a professor wearing shorts! At first, I wasn't even sure if he was a professor.

Everything was different—but all my adventures in India had prepared me to go with whatever America threw my way.

Making My Way in South Carolina

Clemson University was a fabulous place with great facilities, a long list of good teachers, and great support staff. I recall Dr. Imtiaz-ul-Haque and Dr. Vera Anand having a great impact on my education and sustenance there!

My first step was to find a professor at Clemson University and see if I could land some kind of teaching or research assistantship—or just about anything that would provide me with some income. The tuition for the graduate school back in 1996 was $16,300, and I had a total of $1,500 to my name. But if you had a teaching or research assistant position, tuition was reduced from the out-of-state rate to the in-state rate (or, if you were already in-state, a substantial reduction from that). In addition, the assistantship would pay $373.94 every two weeks, and out of that I would pay for tuition, my apartment, and my bills and groceries.

As I was exploring the mechanical engineering department, trying to speak with different professors about assistantships, I saw two students standing in the hallway. I went up to the first student (Arulmurugan) and asked, "Hey, are you new here?"

"Yeah," he said.

"Are you looking for a place to stay?" I asked.

"Yeah, I am," he responded.

"Do you want to be my roommate?"

"Sure," he said. "Sign me up. Also, do you see that other guy standing over there? I think he's looking for a place too." His name was Srinivas. So, the three of us became roommates—that was one thing checked off my list!

Thankfully, talking to the professors was equally productive, and I landed an assistantship within my first week, which was a great relief. For that $373.94 every two weeks, I managed to get an apartment, pay my tuition, and travel across the United States every weekend—which I did in a $500 car I got from my friend that had no reverse gear. Believe it or not, I could parallel park that car just by studying and using the grade of the road!

As a graduate student in a new country, I loved to travel. My first year at grad school, I impressed my advisor, Dr. Harry Law, by scoring 116 out of 100 on my very first test in controls engineering. I hadn't even realized solving some extra problems would give me extra credits! He offered me a position assisting him on some projects, which I did—but every weekend, I would travel somewhere. I put more than 40,000 miles on my hand-me-down car in fifteen months. It was terrific—but it did not have a positive effect on my studies. My grades started slipping, and soon I was getting Bs and Cs instead of As.

And that wasn't my only challenge: I needed to figure out how to fund my love of travel. The only option was to get a job off campus that paid cash. So, I ended up working at the mall selling jewelry, which taught me a lot about customer interactions, sales, and commissions.

It also helped that I shared a deep camaraderie with several of my undergraduate friends from Manipal Institute of Technology, many of whom had come to the States. They were scattered across the country, which meant I had automatic room and board just about anywhere I wanted to go! Shyam was in Florida, Ramnath in North Carolina and later Berkeley, Dinesh was in District of Columbia. When I wanted to see a city, I would go camp out with whatever friend lived there. This is true to this day, when I have friends like Naresh in Wisconsin, Ajay in Bangalore, and Sandeep in Canada.

I loved learning about America and its history. I was so impressed by how good the roads were and how everything was nicely and neatly marked and organized. There always seemed to be a process and a procedure for everything, and—especially after my experiences with corruption and bureaucracy in India—I really appreciated that.

I was intrigued by India and the United States' shared history of liberation from British rule. India won its freedom in 1947, and I was in the States less than fifty years later, so I was attuned to that commonality. I was also attuned to the civil rights movement in the States. I appreciated that the Civil Rights Act of 1964, which ended segregation, had been passed a mere thirty or so years before I landed in the States. I probably would not have been able to come to the

States, let alone go to school here, without the great work of people like Martin Luther King Jr., John Lewis, and Rosa Parks.

I also appreciated late-night comedy. I loved that comedians here could poke fun at the political establishment without any repercussions. You couldn't do that kind of thing in India, where there is a strong culture of respecting elders and authorities—you never say anything bad about them. But here, I loved the humor. Jay Leno became one of my favorites, as did Jon Stewart.

And I loved buttered popcorn! Those were two of my favorite things in the States: buttered popcorn and late-night comedy.

As I observed the United States closely, I was also struck by how intelligent the founding fathers were. As I traveled across the States, I saw that each state was divided based on man-made lines or along a natural boundary—unlike India, where the British divided the country up into states on a linguistic basis. This is part of how the British maintained control for so long: with those linguistic divisions, it was easy to pit one group of Indians against another. In the United States, the founding fathers did things differently, which allowed for a more united population and a more perfect union. Everything here seemed to be the opposite of the British way: people drove on the right, ground floor was level one, switches were flipped up instead of down to turn a light on, words were spelled easier—no "u" in "color" and "metre" was "meter"—and forks were on the right-hand side. The formality was less, and the air was light! Working hard was rewarding. I loved it.

As I came to understand how the US government works, I grew to appreciate the separation of powers between the executive, legislative, and judiciary branches. These ideas gelled in my head, and many years later, when I formed the Caterpillar Asian Indian affinity group at Caterpillar (which now has a sprawling global membership), I drew on these principles to form the leadership structure for that group, including the election of officials and the separation of powers.

The Final Hurdle: Graduating from Clemson

When it came to school itself, graduate school was very different from college. The emphasis was on education and research, and for the first time, the focus was on studying something I was interested in and passionate about—for the sake of learning rather than just getting a degree.

The system was also very different from what I was used to. For example, in general, colleges in India never had homework. Instead, you had three exams called sessionals during the semester, plus a final exam at the end. While the three sessionals' total were given, they took the best of two, which counted toward your final score. Homework was generally not common or a part of your grade at all. Attendance was! One had to maintain a minimum of 80 percent to be able to sit down for the final exams. So, when I started my courses at Clemson, I had no idea that homework mattered. I consistently bombed my homework assignments—until I learned that it counted toward my final grade!

I made my way through my studies with relative success—but getting out of grad school was an accomplishment in and of itself. When I was in my last semester with only six months before graduation, I unexpectedly received a letter from my advisor saying, "You are not doing enough research for me, so I'm not going to be your advisor any longer." This was completely out of the blue; I was doing enough research to qualify as a PhD!

There used to be a joke in grad school that the life of a grad student comprises only four things: apartment, department, advisor, Budweiser. Your advisor is like God. If your advisor says jump, you jump. And if your advisor becomes your adversary, it presents a huge challenge.

Moreover, there was a rumor about this professor that he would keep international students longer than students from the United States. I don't know whether this was actually true, but it certainly seemed to be: my peers and his other students from the United States did seem to graduate sooner than international students.

I was ready to get out of grad school, and this turn of events was a catalyst. Receiving this letter was a big blow, but rather than let it defeat me, I decided to do something about it. I started researching the school's constitution, how the institution was set up, and found I could appeal the decision. I went to the dean and said, "I want to formally appeal this decision. I'm going to convert my graduate thesis to something else, and I want a new advisor, because my advisor is stalling my graduation." I invoked the school's constitution and a number of the provisions it laid out to support my case.

The dean formed a panel, reviewed all my research, and concluded that I had indeed done enough research and should be able to graduate. Rather than putting me back with another advisor, they asked me to start taking a couple of new courses midstream in the semester, meaning I missed several tests. As a result, I got a B and a C, but I was able to graduate—two months sooner than I originally thought I would.

I learned very early on to fight for myself and do what is right. And this experience showed me that unlike India, where there were so many roadblocks and complications, there was always a system in the United States. You might have to research to find out what it is, but when you find the right procedure, you can accomplish what you need or want to do.

I hadn't originally been planning on walking across the stage to get my diploma; that kind of pomp and circumstance did not really interest me. But after this battle, I made sure I was there for my graduation walk, and it was a very happy day for me.

CHAPTER FOUR

Into the Real World

Turning Interviews into Revenue

I had been so immersed in simply trying to get out of school that I hadn't really thought through what I would do next. I also didn't realize that as soon as I finished, my teaching assistantship would end. So, when I graduated a couple months early—in June rather than August—I lost my source of income two months earlier than I had planned. I had no money at all. None. Thankfully, I was able to connect with a professor, Dr. Vera Anand, who was writing a book and offered me a little bit of money to do some editing for her. But it wasn't enough—only $150 or so every two weeks.

Meanwhile, I was looking for jobs, which led me to come up with a creative idea to turn interviews into revenue. I started applying for positions, some of which I knew I didn't want. When I got called in for an interview, companies would offer to fly me out, and I would say, "No, don't fly me. I will drive there." Why? So I could get reimbursed for the mileage at the Internal Revenue Service's (IRS) rate of 32.5

cents per mile! I would drive 200 or 300 miles one way there, get the mileage reimbursements of $160–200, and live off of that money.

I was down to nothing, not a penny, living solely off those reimbursements. I had even vacated my apartment and moved in with my neighbor, sharing his bedroom. I did this for a good three months, until I was finally offered a job I wanted.

In fact, I ended up with potential opportunities at three companies: Goodyear in Akron, Ohio, offered me a job involving research on tires and dynamics—that was moderately interesting; Volvo down in Greensboro, North Carolina, offered me a job relating to testing their trucks; and Caterpillar, in Peoria, Illinois, offered me a job having to do with vehicle dynamics and development of engines and power train systems.

Out of those three, I liked Caterpillar the best. It sounded like I would have the opportunity to work on many different things rather than just, say, tires. But I was also getting pretty desperate for a job. All three said they would give me an offer letter, and I was pretty sure I would have to take whoever offered me a position first. Thankfully, the first offer letter came from Caterpillar—and I took the job.

Will I Make It to Peoria, Illinois?

Caterpillar wanted me to start on July 27, 1998, a Monday. I took everything I owned and packed it into my $500 Mazda MX-6 with no reverse gear. Being a poor grad student, I always outfitted my car with cheap $10 tires with no balancing albeit all top name brands—Pirelli, Yokohama, Dunlap, and Goodyear. With those tires, a car with no reverse gear, and no money, I started the eight-hundred-plus mile journey from South Carolina to Caterpillar headquarters in Peoria, Illinois, on a Sunday, one day prior to my start date. My bank account was down to $0.91, but Wachovia had a $200 overdraft protection, so I knew I could use my debit card to get my final destination.

Within the first fifty miles, one of my tires blew out. Of course, I knew no one, and everything was closed on a Sunday—and I literally had no cash. So, I pulled over on the highway and put on the spare donut tire. I noticed a warning on the spare tire that instructed going no more than fifty miles per hour and to use it only for short distances. But I had no choice, so I carefully drove down the highway doing just fifty miles an hour. Everything seemed to be going alright, so I crept up to sixty, then sixty-five, and soon I was back up to full highway speeds. I drove the rest of the way to Peoria on a donut spare tire. Thankfully, it held up!

I got to the hotel in Peoria in which Caterpillar was putting me up around seven or eight at night, ravenously hungry and with no cash. I only had the free, temporary checks offered by the bank when you open an account. I would regularly ask Wachovia for these because I couldn't afford (and was too frugal to pay) $20 for a book of real checks. As I drove around looking for something, I spotted a Papa John's on Sheridan Road on my drive through town. I was about to go and place an order when this voice in my head said, "They're not going to accept temporary checks!"

So, I made my way back to my hotel room and called that Papa John's instead, ordering a pizza for delivery. When the delivery guy showed up, I gave him a temporary check, said, "OK, thank you, goodbye!" and shut the door before he could protest! I ravenously gobbled down the pizza. For the next five years, every time I called that Papa John's on Sheridan Road, they would say, "Hey, Mr. Sri, just letting you know that we don't accept temporary checks."

My first week of work, I lived in the hotel Caterpillar provided and drove to work every day. One day, my boss asked me where I was staying. "Oh, in the hotel," I replied.

"But where is all your stuff?" he asked.

"Oh," I said, "it's all in my car."

My boss burst out laughing. "So, you're telling me that everything you own in life is in that car parked outside?" he asked. All I could do was shrug and nod. Eventually, I did move into an apartment, but I had no furniture. I had to put my spare tire on my living room floor just to have something to sit on—and I slept on the carpet!

It was a very humble beginning to my life in Peoria, but my job at Caterpillar turned out to be more than worth it.

Life at Caterpillar

My first role at Caterpillar, dealing with the dynamics of machines, was very interesting. Caterpillar makes many different machines but had very few engineers allocated per machine. For example, Honda has twenty to thirty engineers for every model. Caterpillar, on the other hand, had one engineer for every four models.

As a result, Caterpillar engineers are exposed to many different types of machines. The team I worked for was called power train dynamics, and its responsibility was to analyze vehicle dynamics and design control systems for all Caterpillar products—control systems similar to the system that shifts gears in an automatic car.

This position taught me a lot about mechanics, machines, dynamics, equations, modeling, and real-time simulation. It required a combined understanding of physics and electronics and how to make them work together. Before you can design a control system, you have to understand the basic physics. There were only a few people who understood all three elements—the physics, control logic/theory, and the electronics—and I was excited to be one of them. It was a terrific role, and I truly enjoyed it—so much so that the whole plan for my future shifted.

When I left for grad school, I thought I would come to the States, figure out how to do business, maybe start a company, save a little bit of money, and then return home to India. When I came to Caterpillar, I thought, "I'll be here for six months and then see what happens." But the roles I had were so interesting and so technically challenging that I kept extending my stay. I often seemed to be handed the tough roles and assignments that nobody else wanted, and that kept me going. I went from a power train dynamics engineer to a test engineer who tested the control logic to a team leader for new product introductions. I became a Six Sigma Black Belt. And later, the head of Cooling Design Center—a division responsible for all cooling systems in all engines, including the specialized marine engines—recruited me to work in that division.

Caterpillar made a full series of marine engines for large boats and yachts, from 7 to 175 L. It was quite a profitable division for the company. In fact, they made better profit margins on the small quantity of marine engines they produced than on all other captive engines in other products.

But there was a problem: the engines were, to put it bluntly, not faring well. In fact, they were failing all over the world. The issue was the cooling system. Marine engines are essentially captive machine engines with a few components upgraded to "marinize" it. A high power-to-weight ratio was crucial for marine engines. A truck engine, for example, produces only about 300 hp. By upgrading a few components such as the turbo, rings, and aftercooler, the same engine could be converted into an engine that produced 1,000 hp—three times as much.

One of the key components that enables this conversion was called a Charge Air Cooler (CAC). To generate more power, one needs more fuel and to support it with more air. More air can be packed into the cylinder heads by substantially cooling it, as colder air is denser than warm air. Thus, the air from the turbocharger would be chilled in the CAC using seawater as a coolant. But the CAC component was highly unreliable, suffered from corrosion problems, and would often break down. All it took was a few drops of water leaking to hydro lock the engine and cause catastrophic and spectacular failures. Caterpillar had terrible challenges on the entire range of marine engines.

By this time, I'd done some work in other divisions solving some fairly interesting problems, and the head of the marine engine division thought I could help him solve this gargantuan issue—a problem so large it had already resulted in a class action lawsuit. "I don't know anything about cooling systems or marine engines," I told him, "and the closest I've come to cooling systems is when I supercharged my bike and upgraded the oil cooling systems in that motorcycle."

"That's OK," the head of the division said. "I need your problem-solving skills here." Little did I know just how big a problem it was. Once I joined the division, every week, the field service engineers would throw a stack of paper an inch thick on my desk—the list of all the Caterpillar marine engines that had failed that week around the world. I had responsibility for all the engines from 7 to 43 L, and there were problems with every single one of them—except, of course, for the engines that weren't in production yet. Those had no problems!

Caterpillar had been furiously trying numerous solutions—some new, some old. Some suggestions came internally, and some from external suppliers. It was a mishmash of approaches, designs, and problem-solving. We'd been trying things for years, and it wasn't getting better. It was a quagmire, and now Caterpillar had a class action lawsuit on its hands because of all the engine failures. Caterpillar was a great engineering company, very vertically integrated, but for whatever reason, we were relying on numerous suppliers to help solve this problem. We just kept going from supplier to supplier. Finally, I went to my boss and said, "Look, we are a great company; we've been around for one hundred years. We have all these PhDs and scientists and engineers, but we simply don't have the skill set here to solve this tough marine problem. We just have to admit that—and bring in experts from outside the company. I need your permission to do that."

"Fine," my boss said, "do whatever you need to." So, I went looking for someone with expertise in marine corrosion. I found an individual named Hank Preiser, who had worked in the US Navy for almost fifty years and had dozens of patents on cooling systems. Hank was seventy or seventy-five years old at the time and long retired, but when I explained the problem, he said, "This sounds like fun. Maybe I'll consult!" I then found two other experts in the United Kingdom who had worked on submarines there and brought them all together to form a little SWAT team. Working methodically, we determined that the problem was not specific to the component but was, in fact, systemic. This marine issue was particularly interesting and challenging because it was a result of a collage of seemingly disparate issues— from the design to how they were being built to how they were being tested, all the way down to how the engineering teams were organized.

People would often refer to marine corrosion as a "black art" or voodoo—for good reason. Sometimes things worked, sometimes they didn't, and no one could quite tell why. There was lot of anecdotal evidence; for example, I discovered that ships that docked in New York had many issues, while ships sailing in the middle of ocean seemed to have fewer issues. It turns out the New York Harbor was so polluted with heavy metals that the seawater brought in to cool the CAC was causing corrosion, whereas seawater in the middle of the ocean was actually quite clean and, therefore, did not corrode the

CAC. Also, there were fewer cycles on a seafaring boat compared with a boat operating at the harbor.

As I continued to investigate, I came across other parallels and anecdotal observations. For example, if you buy a new car in the south and drive it around down there, and then take it up north, it won't rust as fast. This is because it's had time to slowly gather an oxide layer. Meanwhile, if you buy a car in the north and expose it right away to a season of snow and salt, the body will rust more quickly. It turns out, if you take a shiny piece of metal and shock it with dirty water, it rusts faster. If you give it time to build up a nice oxide layer, it lasts longer.

This, it turns out, was also one of the many factors behind the engine problem: when we built the engines in Peoria, we would test them down at the end of the line, with dirty water. Without realizing it, we were shocking the engines with dirty water right there at the factory.

We saw, end to end, what had to be done, but there was another problem: a human organizational problem. The team itself wasn't organized in a way conducive to either designing successful systems or problem-solving. Each engineer was responsible for a specific component—pumps, CAC, engine heat exchanger, intercooler, and the like. When seawater runs through each of them, it invariably binds them as a system, and therefore, any problem-solving approach should be systemic. However, each engineer would protect their own component and try to make it better without much regard for other components. For example, the main engine heat exchanger was converted to titanium which made it robust but, unfortunately, more "noble" in a galvanic chart. This caused another component upstream to become anodic and become a sacrificial device, exacerbating corrosion. My solution was to reorganize the team of engineers, use a systemic approach, and leave no stone unturned. Deliberately and methodically, we managed to solve this massive problem.

The Search for Purpose in Life

While I found this work interesting, and greatly enjoyed all the roles I had at Caterpillar, I still didn't feel like I was doing what I truly wanted to do, what I was truly meant to do. I didn't feel like I had found my calling, my true purpose in life. So, I set out to figure out what that was.

Peoria is a small town, and I had a lot of free time when I wasn't at work, which I filled with exploration. While I was working at Caterpillar, from 2000 to 2005, I was experimenting to see what I really liked doing, what I really wanted to do with my life.

In 1998, when I started my job at Caterpillar, I acquired my very first internet dial-up. I was fascinated by NASDAQ and Dow Jones and how their stock prices were digitized. It got me thinking, "Why wasn't the Indian stock market digitized?" I knew nothing about stocks. I didn't know how to make money on the market. But I had an idea: I would make every stock price in India visible on the internet. I created a site called merastock.com, which translates to "mystock.com." I managed to engage a broker in Mumbai, who would email me a text file of all the closing prices every day, and every day I would upload them to my site. For the first time, anyone in the world could instantly see what stock prices in India were, without being on a trading floor.

This was my very first company in the States, and I was able to start it practically overnight. I saw how easy it was to start a business here compared to India, and it became clear to me that the United States was a better place to do business. But my heart was not in this stock company. It wasn't my calling. So, I kept exploring.

During one visit back to India, I ran across my old friends from the granite industry. They were the people I had made the rock-cutting jet burner for, solving their most challenging issues. They convinced me that the United States was a big market for granite. So, I decided to bring some granite samples back to the States. Polished granite was not only very aesthetically appealing, but it was also durable and felt natural—connected to the earth and sustainable. The processing of granite also used no chemicals, a fact that was very attractive to me. So, I filled up two 32 kg suitcases (142 pounds) with over fifty pieces of granite samples, each about two inches by six inches by one-quarter inch. I knew each of the samples by their trade name: Diamond Black, Colombo Jubrana, and the like. Of course, I didn't know anything about selling granite, just mining it. But I contacted a few builders and discovered that there is a good deal of money to be made selling granite. However, after exploring the granite-selling business for a bit, I found this line of work didn't speak to me. I wasn't passionate about it. So, I gave it up—and ended up using all the sample pieces of granite I brought back from India as shims for my furniture!

After I left my granite-selling business, a doctor friend of mine told me there was a huge nursing shortage in the United States. "You're an enterprising guy," he said. "I see you working for the community. You and I should start a company placing people."

"Sure!" I said. And, like I always do, I immediately went at it with full force. Our company, which we named DiscoverRN, helped institutions find registered nurses. I even had a partnership with the local government in Tamil Nadu, India, where they helped train and place nurses. Through all of this, I discovered that I was very good at quickly finding people and resources—a skill that would come in handy numerous times in the future. But again, my heart wasn't in this business.

For a while, I volunteered at dog shelters and learned how to train dogs. Given my farm upbringing, and how much I love animals, I thought maybe I should be a veterinarian. I looked into studying to be a veterinarian at Purdue but quickly discovered how challenging it was—and how complete a shift it would be from what I was doing—and concluded it was not for me.

I returned to my love of planes and signed up for flight lessons— one week before 9/11, which obviously brought my flying lessons to a sudden and complete halt. After that, I did not pursue my dreams of flight any further.

Around 2002 or 2003, one of my friends said, "We should start a dollar store; it's a great business." So, I learned all about potential store locations, traffic flow, how barcodes work, and inventory management. But once again, my heart was not in it.

In 2004, while I was exploring lighting for my house, my commitment to LED lightbulbs returned to the forefront. Since my college days, when I hauled around an armful of blown fluorescent bulbs from dorm to dorm, I couldn't stand how wasteful incandescent lights were. I believed that every home in America should have LED bulbs. So, I started a company called Nonstopled.com with my friend Naresh Bharadwaj, where I sold PAR30 and PAR38 LED lightbulbs. My bulbs consumed a lot less energy, but they had a flaw: I couldn't figure out how to make them dimmable. Although I usually love solving a problem, I didn't feel a burning desire to pursue this—so I knew this wasn't my calling either.

I was also very involved in the community in Peoria. This exposed me to many different people and led me to take on leadership roles, including serving as editor and later vice chair for the central Illinois Society of Automotive Engineers (SAE).

Along with my community in Peoria, I wanted to maintain my connection to India. I knew that whatever I did, I would always have a strong connection to India. Whatever company I started, I wanted there to be an Indian component to it. Maybe I'd have a supplier in India or a branch of the business there. Even today, I'm working to see if I can have some of the components for Intellihot manufactured there. I'm always thinking about giving back to my home country. Currently, our control boards, the main brains of our units, are sourced from India!

To help sustain my connection to my home country, I joined the India Association of Peoria, which was formed back in 1964 to serve new immigrants coming into town, host cultural events, and keep the Indian community connected. Through my involvement with the association, I would meet the person who would help me find my true purpose.

The India Association

When I joined the India Association, one of the first things I noticed was that while the Indian community in Peoria was by and large well-off economically, they were not very engaged in civic discourse or the political process. So, I started organizing events to engage with city leaders, and the next year I ran to become president of the association. I only had one opponent—an older gentleman who had been in the community for a long time—but I managed to get elected and become president despite having lived there for less than two years.

I became president of the association in 2001, representing the community. When 9/11 happened, Sikhs and Punjabis in this country started being attacked, because people mistook them for Middle Easterners on account of their turbans. Two weeks after 9/11, I met with the mayor of Peoria, Dave Ransburg, and he proclaimed an Indo-American Friendship Day in Peoria to make sure our community wasn't subject to that kind of violence.

There was no social media back then to connect people. So, the association used to publish a directory every year of all the Indians living in Peoria, along with a newsletter to everyone sharing what was happening in the community. This kept us all connected with one another and with our community.

While I was leading the India Association, I hosted many new people who joined the community, helping them find apartments and houses and onboarding them in their new city. I also increased our engagement with the city and added a sports committee and a youth committee. These committees would frequently put on events for association members.

At one of these events, I noticed a guy I hadn't seen before, wearing somewhat geeky-looking glasses, standing in a corner. It was clear he was new in town. As president, I always tried to engage with new people, ask how they were doing and if they needed any help as they got settled. I did this because I knew from experience how welcoming it was, especially as an immigrant, to have people offer assistance; tell you where the stores are, tell you where to look for an apartment, or offer anything to help you get on your feet in a new place.

I learned that this guy was new to town, had just graduated from Florida State, and had just started a job at Caterpillar. Sensing a connection, I asked, "So, what do you do in the evenings?"

"Well," he said, "I usually just go home and play with my robots."

"Oh," I thought. "I need to be friends with this guy."

That man was Siva Akasam—who would become my best friend, my partner in crime, my coinventor, and, eventually, the cofounder and chief technology officer of Intellihot. In all of that, he would help me find my true purpose and calling in life.

CHAPTER FIVE

Siva Akasam—Cook, Creator, and Master of the Prototype

S iva was born in India, like me, and went to school at the Indian Institute of Technology (IIT) in Mumbai (Bombay)—one of the most prestigious schools in the country. In India, saying you went to IIT is like saying you went to MIT here in the States. Like me, Siva came from a middle-class family—not poor but not wealthy. If something breaks, you don't just buy a new one, but you have to fix it—whether it's your scooter, clothes, a lightbulb, or a radio. So, from an early age, Siva was fixing things. In fact, one summer vacation, he and his cousin fixed somebody's radio, and the next thing he knew, people were coming from all over, bringing them devices to be fixed.

Siva's father was also an engineer—a civil engineer—and Siva had a natural interest in mechanical engineering. He had a core curiosity about how things worked. He went to school for mechanical engineering, and though his grades were not the best, he managed to make it through in the usual four years (although, he says, if you talk to his professors, they might have some colorful things to say about him!).

Siva's interest in robots stemmed from an interest in how to take electrical energy and convert it to mechanical energy to produce useful work. At IIT, students complete a thesis, and Siva took on the fairly large task of building a serpentine robot designed to aid in search and rescue operations after earthquakes. When a building collapses, there are often very tight spaces and corners that a person can't fit through or go around. This robot was designed to go into those spaces to investigate whether it is safe or hazardous and whether there are people there in need of rescue, without the human responders being endangered.

That was Siva's entry into the world of robotics. Robotics is a mix of electrical, mechanical, and control software. They are thoroughly intertwined; there is no single line demarcating one from the other. That's why, even though he trained as a mechanical engineer, Siva progressed into electronics and software. He continued his work with robots and control systems in his master's program at Florida State University—coming to the United States for his graduate degree, as I did. He had some terrific professors and completed his master's degree in two years.

Like me, Siva found many things to love about the States. One of those things was how quickly he could get the parts to build robots. In India, it would take weeks or even months to get a part; in the States, he could order a part, and it would arrive in two days. In India, it was a long and arduous search to find electronic components; in the States, he could go to RadioShack and find just about anything he needed. It was paradise.

Siva's initial plan was to do a PhD and then go back to India to teach at IIT. He felt there were students there who, like him, would benefit from a different way of teaching. But after two years of graduate school, he decided that while he enjoyed teaching and coaching on a limited basis, doing it as a full-time career would not be his cup of tea. So, instead, he decided to look for a job—and got hired at Caterpillar.

At Caterpillar, Siva worked on what is known as a More Electric Truck and then on hybrid trucks. His team introduced a technology now called "low code"—a software development approach that requires very little coding. In low code, you draw block diagrams, and the program generates the software to simulate the operation, all virtually. So, the team would have a model mathematical equation for the truck, the controls, and the drive train and could see what changes they needed to make to the configuration and try out different options—all without spending a dollar on hardware.

From there, Siva moved to working on autonomous trucks, specifically for mining. Much of that mining work, although it pays well, is very monotonous and very hard on the truck operators, and there is a high rate of attrition. The initiative Siva was working on was to automate that whole part of the mining process.

Siva's team would create virtual mines and virtual roads to develop trucks that would drive by themselves. The truck would receive a mission to go to a certain quarry, and it would drive there by itself. Autonomous mining trucks are now in production; if you hear about autonomous mining in Philadelphia or Australia, Siva and his team played a part in that.

When the Australian mining customers first came in to check out the autonomous driving systems, Siva's team was not ready to present the concept. But the customers were there, so the team pulled up the PowerPoint and began presenting—until the top executive of the mining company who was present said, "I don't want to see a Power-

Point," and started pushing random buttons. Siva and his team were sweating bullets, not knowing whether it would all come crashing down. This executive was the decision-maker, and this was potentially a deal worth hundreds of millions of dollars. But after thirty seconds of the executive pushing buttons, the system did not break.

"Alright," the executive said, "now explain to me how this works."

This was one of the first big customer interactions Siva had, and he immediately saw how ease of use for the customer is a critical element for success. You can have all the fancy technology you want, but when it is presented to the customer, all that matters is what the customer sees and perceives. You need to abstract what the customer really needs and be able to present it quickly and simply. You can do ten different things in an experiment, but what is of real value to the customer—whether that's cost, environmental impact, or efficiency—must be very, very clear.

Along with virtual demonstrations, Siva's team would go to proving grounds in Tucson, Arizona. There, in 110-degree weather, they would sit in the cab of a truck with a customer, sweating away, and then say, "Now, imagine if this was autonomous. You could sit in the office, in the air-conditioning, have a cup of coffee, go to the bathroom, and not have to worry about this."

Along with all of the engineering knowledge, many elements of business that Siva and I learned at Caterpillar—customer interaction, talking to vendors, building a team—have become a part of how we operate Intellihot today.

Siva also valued his team members, who were, as he puts it, "off the chart smart." Siva felt that Caterpillar did a great job finding brilliant people but didn't do a very good job deploying them. A talent pool like that could do wonders, but Siva felt the job took their talent, inspiration, and ability and throttled it. The team would propose configurations he knew would work, but the proposals would never go through. The company would spend millions on something the team knew was not going to work, but they wouldn't spend $100,000 on something his team felt would be useful to the company and to the customer. Siva felt there was more to be done, and that is a big part of why he ended up quitting Caterpillar after ten great years.

A Perfect Team

The more I learned about Siva, the more I got to like him—and vice versa. Siva told me he liked playing with Mindstorm, a robotic set that Lego put out. In fact, Siva's first paycheck from Caterpillar went to buying Lego Mindstorm. I told him about Handy Board, a handheld robotics controller developed at MIT. After that connection was made, we started hanging out and playing around with our various robotic projects almost every day.

Siva also had an interest in sports cars and had a Chevy Camaro—which he always drove very fast. He smoked like hell and listened to heavy metal and collected the bottles of all the different kinds of whiskey he sampled. He was my kind of guy.

We were both single at the time we met. I lived a pretty spartan lifestyle; I only had maybe four dishes. But Siva only had two dishes and two mismatched bowls. I thought, "Man, this guy has me beat!" I ended up bringing some plates and silverware to his place just so we'd have enough flatware to eat a meal together! Despite this lack of dishes, Siva was an amazing cook. In his single cooking vessel, he would make all kinds of terrific authentic-tasting Indian dishes. In fact, he was so good, I would often charge him "Baingan bhartas," a wood fire–roasted eggplant dish that he made to perfection, as my fee for doing any tough tasks!

Siva and I never worked in the same group at Caterpillar. I did a lot of physics-based modeling there, and I had a basic knowledge of the electronic side. Siva was the opposite: he knew a great deal about the electronic side and had a basic understanding of the physics. This combination of knowledge worked seamlessly: he could talk about control strategies and I could understand it, and I could talk about the physics and he would understand why certain things should be controlled. We made a good team; our areas of expertise complemented each other perfectly.

We started out working on our little control boards. Siva had developed a smaller version of the Handy Board, which we used to create some experiments and build some small robots. Then he started using a Microchip Technology–based PIC 16F876 microprocessor (a very popular microprocessor at that time) used in all sorts of equipment, such as robotic controls, process controls, microwaves and ovens. It had a set of inputs and outputs you could configure to read digital and analog signals.

Siva quickly learned how much I loved motorcycles and frequently helped me work on my bike, which led him to an interest in motorcycles as well. I was, as I had been since childhood, a very hands-on experimenter. I had a toolbox and was always tearing apart my motorcycles. At first, Siva wasn't really into taking apart engines or bikes, but as we started working together, he became more and more interested. He bought a used motorcycle and then another motorcycle, and we'd be up until two or three in the morning trying to tune our bikes and make them better. Now, he's probably better at taking a motorcycle apart than I am!

At one point, Siva and I had five motorcycles between the two of us—although only two would be operational at any given time. One day, Siva took his motorcycle to a racetrack and soon found himself with a group of semiprofessionals doing donuts around him. "I don't have the skill to ride like them," Siva thought, "but I think we could improve our riding skill by making some adjustments to our bikes."

We wanted to be really terrific riders, coming out of curves fast, with reflexes like a professional. But we were simply not at that skill level, nor did we want to kill ourselves learning. We needed to brake faster; we needed to accelerate faster; we needed to shift gears faster. We couldn't do that mechanically—our hand–eye coordination and reflexes would never be that of professional racers. But maybe we could delegate that part of it to a control system—and then we could just enjoy the ride. So, we started to virtually design a control system for the motorcycle.

The Smart Motorcycle

At that time, the world was moving toward autonomous machines. We were working on several of those at Caterpillar; in fact, Caterpillar was at the forefront of many cutting-edge technologies. I knew Caterpillar was a great company, but that was not the outside perception. When most people think of Caterpillar, they think of a company that makes large trucks or big tractors; how advanced could that be? But believe it or not, it is quite complex to control and run those machines. Even companies like Boeing would come in to have Caterpillar consult them on their simulation software.

Siva and I would often talk about smart machines and work on all sorts of hobby robotic projects. Initially, we were thinking about making a robotic snowblower—because after our first winters in Peoria, we were pretty tired of shoveling snow! Then we began modeling the physics on a computer using freeware, similar to the way we did at Caterpillar, creating what's called "hardware in the loop simulation."

Here is how it works: imagine a car. Every car has a controller to control the gear shifts, the anti-lock braking system (ABS), and all the automated systems in the car. That controller receives its input from a real car with real sensors. But initially, when that product is developed, you can't put the system in a real car to test it. If things go haywire, bad accidents could happen. Instead, you write a series of equations to approximate the behavior of a car, essentially reducing the car into a physics-based model. You have the electronic controls—the boards and wires—interface with your computer, and you have the physics-based mathematical model of the car in your computer. That physics-based model can produce outputs that look like the real car, and the controller can interface with this physics-based model for the purpose of testing. You can do all the development work on a physics-based model, which replicates the actual car, without having to drive an actual car.

This technology was very state of the art at that time. Using this cutting-edge method, I created a model of the motorcycle I owned at that time, the Yamaha FJ1200. We took the control we were designing for the snowblower, and Siva redesigned it specifically for the bike, with sensors for ground speed and throttle position for shifting the gears. Anyone can open up the throttle on a wide-open road, but curves are tricky. So, our control system was the brain for the motorcycle that would essentially sense ground speed, sense how fast you were coming at the curve, and shift for you—giving you the reflexes of a seasoned rider.

Every day, we'd work on this motorcycle project after our day jobs, often at Siva's house. Here he had a workout bench to the side of the room, and every time we had to brainstorm an idea, we'd go do twenty pushups to get the juices flowing. We would also work in the open carport of my condo, testing out the bike until midnight, sometimes one o'clock in the morning, until the condo president came and said, "You can't be running motorcycles here at one o'clock in the morning. It is too loud!"

I was so excited about this smart motorcycle that I even took the idea to the vice president of T&SD (Technology and Solutions Division) at Caterpillar. Full of that self-confidence instilled in me by my mother, I said, "Look. One day, I'm going to be the CEO of Caterpillar, and I think we are not doing things right here. First of all, we have too many people focused on components without thinking about the larger picture. I think you should send me to study systems engineering at MIT to learn how to improve things here. And I think we should put a demonstration project together to show the world that Caterpillar is not just a trucking company or a bulldozer company. I think Caterpillar should produce something that is out of this world. I think Caterpillar should make a motorcycle." And I showed them a picture of a CAT-branded motorcycle.

He laughed. "OK," he said. "Interesting. You want us to sponsor you to go study this new course at MIT, and you want us to make motorcycles." He waved me away.

Little did I know they would later take my idea and send it to the reality show *Orange County Choppers*, hire Paul Teutul Sr. himself, and produce a Caterpillar motorcycle. And then, to top it all off, they sent the future vice president of the group to MIT to take the course I had recommended.

My journey, on the contrary, took me in a very different direction. While my smart motorcycle didn't take off at Caterpillar, it did unexpectedly prepare me for what was about to happen.

How Hard Can It Be?

It was while we were working on the smart motorcycle that my water heater broke and flooded my basement. Coincidentally, this was the same time I was solving the marine engine problem for Caterpillar. Who knew that my job at Caterpillar working on marine engines and the relationship of my work there to engines and water would have such an influence on what I do today! But there I was, standing in my flooded basement, realizing that I am prepared for this. The combination of working with water-based technology and smart control systems gave me a very unique perspective, right from the moment the water heater broke.

I immediately brought Siva in to help me with this problem. On the weekends, most people sleep in. Siva and I were different: we'd get up at six o'clock in the morning, go to Denny's for breakfast, and talk about what things we could work on. So, when my water heater broke, we immediately started brainstorming how to create better heating and cooling systems.

There are so many different elements that need to be cohesively integrated into the system of a water heater. It's a huge task, and taking it on was pure foolishness and bravery. But our motto, as always, was, "How hard can it be?"

This had always been my motto, from a very early age. When my father got his first car, I figured out how to push start it. He was shocked that I could do that, but to me, it was easy. The exit from our garage had a gentle slope. I found that if I put my back on the front fender creating a V-like fulcrum, I could, with my sixth-grade hamstrings, roll the car backward. The slope helped attain a low but sustained speed, and reverse gear being a very tall gear ratio, I easily started the car. One day in seventh grade, unbeknownst to my father,

I took the car and drove it around. I got a sound thrashing for that! After this, I didn't drive a car again until I got to the United States—but I knew exactly how a car worked. So, when I got to the States, I thought, "How hard can it be?" This might sound like fiction, but when I got my first car in the States, I jumped right in and drove away. I was a bit squirmy for the first couple miles, but I figured it out, all on my own—despite having no formal training in how to drive a car.

When it came to figuring out the water heater, I didn't have any training or background in anything directly related to water heaters. I just had Siva by my side—and that deeply instilled self-confidence. Together, we brainstormed and experimented and tested until we found the solution. Every step of the way, no matter how great the challenge or hurdle we faced, Siva and I would say to each other, "How hard can it be?" And we would keep going. To this day, Siva often quotes this motto back to me, saying, "We started this journey saying, 'How hard can it be?' It can't be that hard. Let's just try it."

Working together, Siva and I quickly came up with the idea of repurposing the control system we had created for the motorcycle. Then, starting that very weekend and stretching over the next few weeks, we built a prototype.

There were a number of bumps along the way. To start, I went to Home Depot, where, ironically, they had a clearance on mechanically driven residential tankless water heaters. Tankless was problematic, and they wanted to get rid of it. The units were $425, and I bought one. When I stripped it down, I could see it had interesting mechanical controls, but I knew right away that it wasn't sufficient to meet the requirements of precise temperature control.

We decided that we were going to wrap our electronic brains around this device. But first, we needed a way to start the flame. So, we tried to buy an igniter—a device that creates a spark to start a flame. I called the local supply house and said, "I want to buy an igniter."

"Are you a homeowner?" the representative asked. I said yes.

"Why do you need it?" he asked.

"I'm trying to build a smart water heater," I replied.

"You're *what?*" he said. "You're trying to do *what?*"

He could not understand what I was trying to do—and, of course, my accent didn't help with the misunderstanding. Between the two, he would not sell me the part. I called two other parts stores, and they all ended the same way: "Sorry, we can't sell homeowners this part."

"Forget talking to these guys," I said to Siva. "They're just giving us the runaround. Let's just build our own igniter. How hard can it be?"

So, we built an igniter, using a regular transformer, and it produced a spark—but we ran into another challenge. Producing a spark creates a lot of electromagnetic radiation—so much that if you have any electronic integrated circuit (IC) chips nearby, such as in your radio or your music system, it will reset the IC through electromagnetic radiation. Now, it happened that Siva had a Casio keyboard in the room next to the room we were working in. Every time we would spark our igniter, the Casio keyboard would reset and power cycle with a start-up chime!

We looked at each other. "This is … a lot of radiation," we said. "What if it affects our brains? What if it turns us stupid?" We considered this for a moment—then shrugged and continued with our work, keeping at it until we had a functioning igniter.

We repurposed a heat exchanger and built controls from parts cobbled together, using bits we'd purchased off the internet and components from other devices—a modulating gas valve, a flow sensor, and an igniter.

I took the control system we had built for the smart motorcycle. A few temperature sensors fed the repurposed motorcycle controls. With these rough pieces, we could build a working prototype. Siva was the one who had written all the software for it, so I asked him to work on it—but he said, "Why don't you give it a shot?" So, I taught myself C code. I was not—and still am not—a programmer, but how hard could it be?

After a few weekends, I had a version of the code working. It would detect water flow, spark, open the gas valve, and carry out the ignition function. The whole process took almost three minutes—far too long to be commercially viable—but that wasn't important at the moment. The point was, the code and control logic worked. I was so proud to have done it on my own that I immediately showed it to Siva. He certainly could have made a better version than I did, but I think he wanted to see how committed I was to this idea!

At this time, Siva lived in a three-story home, and his mechanical room was tiny, essentially the size of a closet—but we would cram ourselves in there to run our experiments with the prototype. We'd run experiments using gas, without really fully understanding how explosive gas was—I hadn't changed much from my college days when I kept gas in my dorm room! Every time the unit would ignite, we would say, "Great, let's put in a little more gas!" Thinking back, there was a very high risk that if something had gone wrong, there could have been an explosion. In fact, a couple of times, we did have a mini explosion. The whole house would shake, and Siva's wife would come down and say, "What was that noise?" We were definitely the archetypal mad scientists in the basement. Thank God for our guardian angels who were probably on double duty around us!

The Prototype

Part of our journey was trying to find the right people to help bring our concept to a commercially viable reality. We knew that the critical characteristics of this new heat exchanger had to be thermally responsive, light, and very robust to corrosion and thermal cycles—all characteristics that don't necessarily intersect easily. It's like trying to build a car that is simultaneously the most reliable, the fastest, the most efficient, and the cheapest: not an easy task. So, we went looking for people whose expertise could help fulfill my vision of a very efficient and smart machine.

We needed to find someone who had the skills and knowledge to design this particular type of heat exchanger. First, I looked at Bradley University, the university in Peoria, and talked to a professor there. He expounded on the theory and slightly berated us for not understanding how tough this challenge was. We quickly passed on him, given his attitude and lack of vision. Also, he wanted a large sum of money—which we did not have.

Then, we found an engineer in Nebraska who was more affordable. His expertise was in designing thermal power plants, but we thought he might have some insight. Siva and I drove out to Lincoln, Nebraska, to meet with him. He turned out to be Indian as well, and after we'd done some experiments, I met up with him again in India and had him sketch out an initial design. But the design he sketched didn't make sense to us. His main experience was in designing heat exchangers for large power plants, and he couldn't quite conceptualize something this small.

However, we took his sketch as a starting point and, through experimentation and trial and error, made some amendments. We also consulted with a few people around the world with experience in the appliance world or working gas combustion products. As I had discovered when I started my nurse placing company, I was fairly resourceful in finding people—a skill that has continued to come in handy to this day!

Overall, we found that no single individual could fully help us get to our destination. It was really going to be up to Siva and me to use all of these inputs as a starting point and create something radically different that met our vision.

We then brought our modified design to my father's factory in India; I had no idea where to find the resources, money, and place to build such a thing in the States. My father is a mechanical engineer, so he understands fabrication. He works in the physical realm, not the theoretical realm. When I first started working at Caterpillar, my father would ask me what I did, and I would say, "I'm figuring out control systems or vehicle dynamics." But I don't think he understood what I was talking about. Unless I was actually building something, to him, it wasn't real engineering.

My water heater project, on the contrary, was a real-world, physical invention—and that he understood. When I started working on it, I explained to my father that while most heat exchanges were created out of copper, I was planning to create this high-efficiency heat exchanger, which also had to be corrosion resistant, out of stainless steel. That very evening, my father ran two experiments. He filled one copper vessel and one stainless steel vessel with water, turned on the heat, and set a timer. Then, he called me and said, "I don't think your stainless-steel idea is going to work—because stainless steel is twenty times worse than copper. It takes twenty times longer to heat water in a stainless-steel vessel than it does in a copper vessel—and your whole idea is to heat water quickly." What he said is true, but I knew it could be overcome.

When I told my father, "We need to fabricate this prototype we have designed," he was supportive and very helpful in having that done at his facility. Fabrication is his area of expertise. You can give him a set of drawings, and he can cut, weld, grind, heat, treat, and build massive structures—as complicated as the large heat exchangers used in nuclear reactors.

At the time, India had a lot of thermal power plants, and they were in the process of setting up nuclear power plants. One, in Chennai, was a new type of reactor, called a fast breeder reactor, which was designed to produce more fuel than it consumed. My father worked on nuclear power plants (as well as thermal power plants), so he had many books on fast breeder reactors. My father also helped set up India's largest press at that time to form the large pieces of steel used in pressure vessels. He was used to thinking about fabrication, so building a relatively small prototype water heater would be no problem.

So, at my father's factory in India, we built a custom prototype based on our first-pass design. It took me another three or four months to ship it back to the States, as I didn't know anything about exports. That was, in fact, the greatest lesson I learned from building this design: how to export from India! It required lots of certifications, a fumigated crate, and the like.

The model finally made it to the States. It was very crude and weighed 170 pounds—but it worked. It ran at very high efficiencies, extracting nearly 99 percent of all heat. As I reflect on the design now, it was advanced for its time, having a down-fired burner and an all-stainless design. Not bad for a first-pass design from two engineers who nearly flunked thermodynamics!

Connecting the Dots

With this water heater project, I'd finally found my calling. I knew it was important. I knew it would be challenging. And I knew it was the right direction. All the dots in my life were leading to this—my love for intelligent machines, my work on control systems, my work on marine engines, my interest in energy efficiency—and, of course, my friendship with Siva.

Siva loves a challenge. He loves trying to figure things out. He loves math. He didn't start out with the focus on energy efficiency I had, but as we worked on the water heater challenge, it grew on him, and today he has dedicated his life to solving global warming. It aligned with his belief that intelligence should be used for good. Some people think of intelligence in an egoistic way, saying, "I am intelligent, so I am great." Siva, on the contrary, is all about using his intelligence and creativity for the betterment of others. "What use is it being intelligent," he always says, "if you can't use it for the good of the world?" I am lucky to have found someone as altruistic and humble as Siva.

Overall, the intrigue and challenge of solving this problem and creating something new were very alluring—for both Siva and myself. It was a triangulation of seemingly disconnected decisions and events that all led to me being in the right spot at the right time. I was back with my true loves: smart machines and energy efficiency.

As we worked on the water heater, Siva would ask, "What is our mission in life?" And I would respond, "It is our mission in life to bring good, clean, hot water to people." And we would chant this like a mantra.

I knew beyond a shadow of a doubt: this was where I needed to be in life.

Inventor to Entrepreneur to CEO

Intellihot was officially founded in my basement in 2006. For the first few years, we were essentially a research and development company, experimenting and running tests in my basement on evenings and weekends. Our goal was simply to figure out how to make our concepts work.

Proximity to our work at Caterpillar allowed Siva and I some free time in the evening. In addition, we could go to work early—6:30 or 7:00 a.m.—and leave early, leaving us more time in the evening to focus on Intellihot. While other people would use their evenings and weekends for leisure activities, we went straight to my basement to work on various contraptions. We had fascinating, challenging problems to solve, and that is all we wanted to do with our free time. Solving hard problems motivated us!

We were still working at Caterpillar in 2007 when we built the first condensing heat exchanger prototype with a down-fired burner in my father's factory in India. When we finally got it back to the

United States, assembled it, and were able to run it, we were ecstatic. "My god," we said, "this actually works!" But this prototype was not something we could take to market and sell—or even something we could take into production. In truth, it was hardly a prototype at all. It was still just a concept, a bunch of elements—engine, igniter, controls—hemmed together. It looked like Frankenstein's monster—and it got red hot!

Siva and I knew that in order to take this into the next phase—a product we could bring to market—we had to put in more than evenings and weekends. To do this, we needed to quit our jobs at Caterpillar and pursue Intellihot full time.

Considering where we both were in our lives, this was not the most obvious or logical choice to make. Siva just had a baby, and I had one on the way. We were leaving good, well-paying jobs—the very jobs for which we had moved to Peoria!—and Siva was even up for a promotion at Caterpillar. Understandably, everyone was skeptical. "Why are you quitting?" people would ask. "You have a child on the way, and now you want to go do something you know nothing about, in which you have no background—how are you going to provide for your family?" Our families were asking us the same questions: "What are you doing? You have good jobs, you're recognized and moving up at Caterpillar. Are you guys crazy?"

But we knew we had something real with this water heater, we knew it was worth pursuing, and we knew that staying at Caterpillar, promotions or not, wouldn't allow us to reach our full potential. As Siva says, "We've been blessed with some brains. If we don't use them to their full capacity—what a waste!" I also knew that smart, free spirits like Siva and me are not meant to be caged. Even my boss at Caterpillar once lamented that I brought so many ideas to better things but my wings were clipped.

So, we convinced our spouses to go along with our plan: to pursue Intellihot full time for six months and see what happens. If it didn't go anywhere, we would find regular employment again. And with that plan, in 2009, Siva and I both officially left our jobs at Caterpillar and embarked full steam ahead on Intellihot.

We began with a clean sheet of paper, starting the design from scratch, developing everything simultaneously—the engine, the controls, the electronics, the display. We found, once we were working on the water heater full time, we moved very rapidly. When you move from the corporate world to working for yourself, you can move at lightning speed. There is no red tape to block your way, no bureaucratic chains of command to navigate in order to get anything done. You can truly move mountains. There was an unmistakable energy, almost electric, and I found myself hardly able to sleep.

After six months, Siva and I were in complete agreement: we needed to continue our work with Intellihot. Things were going well, we loved the work, and we both felt we were exactly where we should be, doing exactly what we should be doing. We could not see ourselves going back to corporate life.

This was it. There was no plan B. We were all in.

Looking back, it is shocking how little we really knew about what we were doing. We had set course to invent a new water heater—a complex gas powered one no less (electrics are much simpler and easier)—along with its entire control systems, including the circuit board design and user interfaces. It was honestly a bold undertaking! We didn't realize the full scope of what we were taking on, or how much time and money it would take. We knew it was a challenging engineering problem, but we didn't really think through how, in order to solve that problem, we would need to be supported by adequate funds and facilities and personnel.

But, as I often tell people, you don't actually need to know everything. All you need is a high dose of self-confidence, and you can do just about anything. That's the only thing that made it all possible: the immense self-confidence I had—not arrogance, just an unshakeable and unquestioning belief that it could be done, that we could figure it out. After all, how hard could it be?

Inventors to Entrepreneurs

It was apparent from day one that running an actual company and producing an actual product was far more work than just being an inventor. Being an entrepreneur isn't for the faint of heart. There is so much to accomplish: having the right product, going to market, raising money, not running out of money, finding the right talent, retaining them, and, most important, proving to yourself that this journey you started isn't in vain.

We had completed step one of the big picture: figuring out what we wanted to do in life, what we were passionate about, what our calling was. And we'd completed step one of the business: inventing, designing, and building a product from scratch—a product we truly believed in. Now, we were at step two: turning that invention into a business. How do we manufacture the product? How do we market it? How do we build a brand name? How do we create a place in the market for ourselves? And how do we do that in an industry that has not seen any change in over one hundred years?

We had moved from "What do we like doing?" to "How can we get this to work?" and "How do we turn this into a business?" In other words, we had moved from being inventors to being entrepreneurs.

This shift was a learning experience. Prior to this, I had led nonprofit organizations, and I had headed up large teams—including teams working in areas in which I had no prior experience, such as the marine engine team at Caterpillar. You have to learn quickly, be humble, and follow your intuition. You have to surround yourself with smart people, help them succeed, and find your own successes in theirs.

We connected with Don Rohn, the Peoria coordinator for Score. org, a service that connects small business owners with expert business mentors. Don's company made the cell phone towers you see typically dotting the highways; he was a great businessman who had scaled the company his father started into many businesses. Not only that—he could also run a sub-five-minute mile, and he was the humblest man you'd ever meet.

We learned a lot of things from Don. He liked to develop a well-thought-out plan and secure the right people and cultivate the right relationships to accomplish the broader goals. Don also had a way with finding about more about any person we would interview for a role. He would get behind the niceties and truly assess motivations. He would elicit answers and information that were surprising and were very helpful in securing the right person. One such find was Rod Harrison, who was Intellihot's CFO for many years and guided the company through thick and thin. In fact, while I was searching for a CFO, Don mentioned he knew the right guy who was a former CFO at his own company but wanted to check with him before he put us in touch. Unfortunately, Don had a fall and injured his head, something he never recovered from and passed away in late 2011.

Even though Don never introduced us, I tracked down Rod and connected with him. He had also heard about us from Don and was equally intrigued, especially since it came from Don. After a few conversations and once Rod had completed his own due diligence about us, he joined the company, part time initially and later full time. He was a terrific operational CFO and would walk the factory and engage with people at all levels. He kept close tabs and guided the company as we built the business and developed our brand name. Rod was so committed to our cause that he even named the tag on his car to "HOT WATR." I couldn't have asked for a more committed individual with a calm demeanor than Rod Harrison. He was a rock at Intellihot. This couldn't have happened without Don's foresight and ability to know people and how they all may fit together as a team.

Another of Don's principles was to always ask for a discount, because you'll usually get it. "You get zero percent of the discounts you don't ask for," he would say. This was very helpful for a company with pretty much no capital to its name! To this day, even when we are cash flow positive, we always ask to see if we can get any breaks on price. We call it "low-cost mode"—and it is not the same as being cheap!

Don is also the one who first encouraged us to actually craft a business plan. "You need to put your thoughts down on paper," Don told us.

"We don't need a business plan," we said. "We've got it all in our heads; we know what we need to do."

But Don convinced us we needed to write it down. "In order to scale up, you need to find a space," he told us. "In order to find a space, you need some money. In order to get that money, you need to know how you're going to spend it when you have it. And you need to have that written down in a business plan, so you can show it to the people you're hoping to get money from."

Money was certainly something we needed. For our first several years as entrepreneurs, Siva and I did not take any salary or pay. We worked for free—and we worked seven days a week, fourteen or fifteen hours a day. But we loved what we were doing so much that we didn't even notice how much we were working. We had a blazing energy flowing through us, like an electric current. I couldn't sleep! I was buzzing with so much excitement! I was constantly thinking about Intellihot.

As my board member Mike Reardon rightfully said to me, "If you're an entrepreneur, you even think in your sleep!" Nights and weekends all blurred into one. The phrase, "Have a good weekend!" meant nothing to me. Siva and I didn't know what a weekend was! The saying, "Do what you love, and you'll never work a day in your life," became utterly true for both of us.

That is not to say we didn't have other obligations in our lives. As I mentioned, when I left Caterpillar, I had just found out I was going to have my first child. I knew I needed to find a balance in which I could devote myself both to Intellihot and to my family. At Caterpillar, I had always been very efficient with my time, but this would require a whole new level of efficiency.

When my child was born, I made sure to establish a schedule where I could do everything I needed to do as both an entrepreneur and a father. My son faced a number of medical issues as a child, and I did not miss a single one of his appointments. I dropped him off and picked him up from day care, according to a very strict schedule set by his mother. I could not be more than a minute late!

As I had done from childhood, I learned how to do more with less: more responsibilities with less time. In fact, I became so efficient with my time management that I found I was working the same amount of time outside home at the office, if not slightly less, than I did when I was at Caterpillar. Of course, once my family obligations were done, I would return to my work.

But the true secret to finding balance came down to one thing: passion. I would never have been able to find this balance if I was not passionate about Intellihot. When you turn your hobby, your passion, into your job, balance becomes easy. The time you would otherwise be devoting to that hobby outside of your job now becomes part of your work time. On weekends and free evenings, instead of spending time on your hobby, you spend time on work—because work is your hobby. While you're drinking your coffee in the morning, you are thinking about work—because it's what you love to think about.

If you want to start a business and be an entrepreneur, don't do it for the money. Do it because you are passionate about it. If you are not passionate about what you are doing, if you do not truly care about it, if it is not your calling, you will never be able to find balance. Passion is what keeps you from getting bored or burnt out. This is why it is so essential to find your calling—even if that means trying many different things, as I did!

Entrepreneur to CEO

The truth is, starting the company was the easy part. Opening a bank account, getting a tax ID—I'd done all of that several times over, and even adjusting to the new schedule of being an entrepreneur wasn't too difficult. Many of my experiences up to this point had prepared me for the role of entrepreneur.

What I was not prepared for was raising money. I had never raised money for an entrepreneurial effort before, and I had no idea how to do it—or how difficult it would be. We had very little capital, so we had to be very frugal. This part of the entrepreneurial phase was all about: How can we do more with less? How can we stretch our financial resources? How can we get creative with what we have?

From very early on, we learned ways to bootstrap. Even before I had quit Caterpillar, I went to my local PNC Bank to seek funding. "I'm starting a company," I told Diane Warbritton, the president of the bank, "so I need some money." I had no idea what I was doing, but she was very nice about it and explained that they needed to see a business plan before they loaned us any money.

Siva and I spent the next few months drafting up a very detailed business plan and then returned to the bank, plan in hand. Once again, the bank asked for more supporting documentation, which we collected and brought back to the bank—but the bank still would not give us the loan. "We need to actually see a good sum of money in your account before we can give you a loan," they explained.

Siva and I, being very creative, begged and borrowed from our family and friends and managed to put over $100,000 in our bank account (with the promise that we would return it all as soon as we had our loan from the bank!). "Look," we told the bank, "we have $100,000! Now, we need $500,000 from you to start our business."

Having exhausted all excuses, PNC Bank finally delivered the news: a resounding "no." "We don't fund start-ups or risky businesses," they told us.

It was clear a bank loan was not the path we needed to follow to get our business up and running. The whole process of going back and forth with the bank took nearly six months. Looking back, I can see that had I followed another path from the beginning, we could have moved much more quickly. However, hindsight is always twenty-twenty—and back then, I didn't know anything about raising money, so my experience with the bank was a lesson I needed to learn.

After our unsuccessful attempt to get a loan, Siva and I realized we would need to fund the project from our own savings. We did this while continuing to experiment in the basement. When we quit Caterpillar to pursue this full-time, we put in some additional seed money totaling $40,000, plus an in-kind contribution via the salary-free work we did for a few years.

At that time, we also opened up a "family and friends round" of investment. We did not raise very much from this—only about $50,000 and then about $40,000 in a subsequent round—but it was enough to get us started. The investment amounts varied from $500 to $20,000, and I am especially thankful to those people who stepped up and supported us right at the beginning of our endeavor. To some extent, it showed us who our friends truly were!

Our very first investment came from Jim Durand, a mentor and colleague of mine at Caterpillar. Jim was in charge of the code of conduct at Caterpillar. We were part of a Six Sigma project tasked with developing better hiring and employee retention programs. At that time, Caterpillar was serving as a training ground and losing talented employees to other companies like Motorola and Ford. Jim and I had many a conversation about the code and its importance, as well as the importance of developing relationships with people of different backgrounds. It came in handy many years later as I wrote the first code of conduct for Intellihot.

Along with capital, we needed more space to develop our product. We knew we were at a point where we could no longer operate solely out of my basement. Luckily, in early 2009, thanks to Don Rohn, we were able to obtain space at the Peoria Next Innovation Lab—a small space, only about seven hundred square feet, but a vast improvement over our basements! We also had access to other resources Peoria Next offered, such as conference rooms, printers, and a network of other like-minded people.

We also needed to formalize the structure of our company. In other words, we needed a CEO. Once again, this is where Siva and my natural skill sets and interests are in perfect alignment. Siva is very good at conceptual thinking. He loves to come up with a concept, develop it, and build the prototype. While I am also able to think on a conceptual level, one of my skill sets is also thinking about the big picture, the overall project plan.

I thought this might make me a good fit to take on the role of CEO of Intellihot—but I did not want to simply declare this so. I wanted to approach this role the way Abraham Lincoln approached running for president: when he was campaigning, he didn't say, "Elect me." Instead of asking people to vote for him, he would say, "May I have your permission to govern you?"

I admired Lincoln for his wisdom and humility, so I approached Siva and said, "Siva, I think we need to figure out who is filling what role in this company we started. I believe I have the right skill set that will naturally extend to those roles. May I have your permission to be the CEO or president?" Siva agreed—and with his permission, I officially became CEO of Intellihot. I am grateful for his trust and confidence.

Once I became CEO, I discovered that I was even more suited to it than I had expected. By nature, I would define myself as an introvert. Other than LinkedIn, I'm not on Facebook or any other social media platform. I don't like speaking about myself or promoting myself. However, I found that as CEO, sharing my vision and my passion came naturally to me. When someone would ask me, "Why did you do all of this?" I wouldn't even have to think; the story and philosophy behind Intellihot would come pouring out.

Until I started sharing my story, I assumed everybody would react the same way I did when confronted with my situation. But more often than not, people were surprised at the origins of the company. "You started all of this because you were upset by how much energy was being wasted by your water heater?" people would ask.

"Yes," I would respond, "of course."

While it took me a few years to fully develop and refine how I talked about Intellihot, I quickly learned that I was good at getting in front of people and presenting the vision and value proposition of our product and company. And the more I did it, the better I got at finding and interfacing with people—suppliers, partners, and customers.

Now that we had a space to work in and our company structure solidified, we just needed one more thing: a market-ready product.

Concept Development and Product Certification

With less than $100,000 in the bank, we embarked on our project: creating a fully functional, fully certified model of our intelligent water heater. Usually, companies will spend millions of dollars to create a fully working, fully certified prototype, but we were confident we could do it with much less. We hired our first employee: Mike Gonzales. He was a chance meeting when a few others I had talked to weren't able to offer their time for engineering design. I'll talk more about Mike later, but this chance setup would prove to be very fruitful—the start of a long and cherished collaboration.

We moved swiftly from having nothing to having a fully developed product, ready for certification, in less than fifteen months. It was a dizzying effort. Even today, I am not sure how we pulled off this stunt, which included the development of a state-of-the-art heat exchanger, sophisticated controls, and electronic hardware.

Most companies, when developing a product, don't develop all the systems themselves. They will outsource a few of these systems to other companies with specific domain experience in that field. For example, for $2–3 million, one company would develop the electronics and controls, another company would develop the heat exchanger, and yet another company, with another few-million-dollar contract, would develop the Internet of Things (IoT), cloud, and app. Obviously, we did not have that kind of money. We had less than $150,000, and we wanted to do it all! We simply didn't have the mentality for outsourcing. Following, as always, our motto of, "How hard can it be?" we developed our own controls, our own heat exchanger, our own electronics, and our own circuit board design. It can be done!

At Peoria Next Innovation Lab, with this small bit of funding and a whole lot of hard work, innovation, and ingenuity, we were able to move at lightning speed. We went from experiments to a fully functioning, production-ready unit—a unit very similar to the current unit we produce today—in less than eighteen months. It is truly mind-boggling!

Mighty Mike

Siva and I didn't do this all on our own. Right at the start, we hired our first design engineer, Mike Gonzales, who coincidentally—and to our good fortune—had just been temporarily laid off by Caterpillar and was looking for temporary work. When we hired him, he was the only person on payroll, as at that point, Siva and I were still not paying ourselves!

When we offered the position to Mike, he told us he would work with us on our "project" for six months. "But when Caterpillar calls," he said, "I'll go back." Fortunately for us, as Mike started working with us, he discovered that he loved the work and the free thinking with no red tape. When Caterpillar called, Mike went back, as he said he would—but only for two months. After two months back at Caterpillar, he returned to us, saying, "I don't think I can stay there.

I'm having too much fun here at Intellihot, in this environment of freedom and innovation and no red tape."

Siva and I knew we had found another kindred spirit. Mike is still with us today as a chief engineer for the company. We still joke that his six-month project never ended, and at any minute, he might return to Caterpillar! Without Mike and Siva, none of the things that Intellihot achieved would have been possible. The three of us made a dream team—innovating and executing in perfect harmony. I don't think we ever worked as hard or had as much fun as the three of us did in those eighteen months. Our lab was like a madman's workshop!

Water heaters are a highly regulated product, because when a water heater malfunctions, it can blow up, destroy property, and even kill people. Therefore, water heaters must be certified to very rigorous standards—and building a product that would meet those certifications would require more money. So, Siva and I returned to a different bank—State Bank of Speer Bank—which was a smaller local bank.

This time around, when the bank inquired about our assets, Siva and I offered up the only thing we had: our homes. We used all the equity we had in our homes to obtain a $250,000 loan from the bank. With this move, we were ready for the next phase: product certification via third parties. The stakes were high: if our Intellihot endeavor failed, we would lose our homes.

When we invited the certifying body to come review our work, they told us we needed to have a production facility. So, we cordoned off one side of our madman's workshop and said, "This is where the parts come in; this is where we put the units together; this is where we test the units; and here is quality control. Taken together, that's our production line!"—all in less than eight hundred square feet.

The certifying body (ETL/Intertek) tested our unit and officially certified that it met regulatory safety standards. We also had it tested and certified to Energy Star standards. We now had a fully certified unit we could offer for sale—and we did it all for under $250,000 with just three people: me, Siva, and Mike. All it takes to do something truly amazing is a few really smart people in a room together, united by a common passion and cause.

Journey to Galesburg and Setting Up the Factory

Now that we had a fully certified, production-ready product, it was time for Intellihot to enter the next phase: production and scaling up. This would, of course, require more space, more materials, more people, and more money. It's always tough to raise money—but it's especially tough for a company that really has no products on the market, just one certified prototype. As usual, our path to finding funding was unorthodox.

As we were working on our initial product model, we entered a Launch business plan competition, organized by the East Peoria, Pekin, and Peoria Area chambers of commerce, the Peoria Next Innovation Center, the Morton Economic Development Council, SCORE, Bradley University's Turner Center for Entrepreneurship, and Bradley's Technology Commercialization Center. After making it to the final round with six other finalists, we were declared the winners and awarded $3,000 to help us with our business.

Later that year, we were also one of six finalists in the Innovate Illinois competition, held in Chicago, and between these two events, our little company found itself in a regional spotlight. Then, CNN did an article about the most entrepreneurial places in the United States, featuring Peoria as one of the best locations to launch a small business—and including a little segment on Intellihot. This attention in the press was all new for Siva and me. We were too cheap to even hire professional photographers to take our pictures for the segment. Instead, we just took each other's pictures, and that's what ended up on CNN!

After these accolades and this recognition in the press, and since we had a fully certified manufacturable product, we started being courted by cities and states all around the Midwest. Morton, Illinois, reached out to us, as did Omaha, Nebraska, eager to bring the production of our clean, high-tech, high-efficiency product to their area. "You have a fabulous product," they would say. "It works, it's certified, it's ready for production—you should set up a factory and provide employment in our city!"

Around that time, I received an unexpected call: Don Crose, the plumbing inspector from Galesburg, a city about fifty minutes west of Peoria. That fateful call would eventually result in us locating our factory there. Galesburg is a small, pleasant town of about 30,000 people, which was once home to the main factory of Maytag. The factory employed around 50,000 people. Later, Whirlpool bought Maytag and relocated production from Galesburg to Mexico. This left the city devastated and desperate for a way to create new jobs and revitalize itself.

"I saw your company featured on the news as the most innovative company in Peoria," the plumbing inspector said. "I think we could use your product to make some of our city buildings more efficient and to provide more efficient—and therefore cheaper—water heating for some of the low-income homes the city owns. Can you come to Galesburg and show us what it is you do?"

So, we took our prototype unit out to Galesburg. When we arrived, we were greeted not only by the plumbing inspector but also by Sal Garza, then mayor of Galesburg—a former Maytag engineer who had caught wind of our visit and wanted to check out our product. When we had demonstrated how the water heater worked, the mayor asked, "Where are you planning to manufacture this device?"

"We're still looking," we answered.

"Well," said the mayor, "I am interested in having more small, nimble, innovative, energy-focused, and green companies make Galesburg their home. Intellihot seems to be exactly what I'm looking for. You should set up a factory here in Galesburg."

"We'd be happy to," we said, "but starting up a factory takes money. Would you and the city be able to help with that?"

"We will look into it," said the mayor. He then directed Cesar Suarez, his economic development director, to come up with a plan and put him in charge of executing it. Cesar was a highly energetic and creative individual. He arranged for meetings with several local banks, highlighting to them the potential impact on the local community. We were fortunate to land an engagement with the local Farmers and Mechanics (F&M) Bank, who were great to work with. Along with them and the city, we developed a plan: a combination of a loan from the city plus the rest of the capital we needed via F&M Bank.

But before anything could happen, a formal City Council vote had to be taken to approve the plan. While this was all in the works, the mayor, Sal Garza, paid a surprise unannounced visit to our lab in Peoria—with a thermometer in hand! He wanted to see if our product really worked. Thankfully, our rag-covered, leaky prototype produced hot water at various flow rates and maintained accurate temperature control. He was satisfied that we had the real deal. Next step: a City Council vote.

Mitsubishi 3000GT to the Rescue and the City Council Vote

The City Council vote was set for Tuesday, July 6, 2010. On that important evening, after a full day of experiments, Siva and I wore our best shirts and ties. We set off to Galesburg, which is normally a forty-five-minute drive from Peoria. It was raining that day, and I was doing about seventy-eight to eighty miles per hour—I didn't want to be late! Somewhere near exit 55 on I-74 West, I noticed the front end of my Mitsubishi 3000GT shimmy a bit. While I was pondering why, in the next instant, the entire car spun 180 degrees and flew off the highway at eighty miles per hour. We went into the median, continued spinning, and the car started climbing up on the other side of the median—straight toward the east-bound side of I-74. I could see the headlights of an oncoming truck, which started blaring its horn. I thought my life was coming to an end.

All throughout this ordeal, both Siva and I were calm—no words or screams. It all seemed to happen in slow motion. Luckily, somewhere along our spinning climb to the other side, the car came to a halt and drifted back into the ditch, now facing the opposite direction of our original trajectory on I-74 West. Siva and I spent a few minutes just taking in what had happened. I managed to restart the car, point it in the correct direction, and drive it out of the ditch. We continued on our way to attend the most important City Council meeting and vote of our lives.

On the way, I couldn't but help ponder why this had happened. Then, it dawned on me: I had just put a brand-new tire on the front right side of the car. While replacing only one tire wasn't smart or recommended, I did not have the money for two new tires. In those days, I drove until my tires were bald! In the rain, the left-front tire—

which was bald—started to hydroplane, but the brand-new right-side tire continued to pump water and maintain contact with the road. When this happened, it spun my car counterclockwise in an instant.

Siva agreed with this explanation: "Let's keep this near-death experience to ourselves," he said, "and not scare our wives!" I was thankful to the sporty nature of the 3000GT and credited it for not tumbling over. In fact, I still have that car and plan to keep it for life as a token of affection for it having saved my life. After all that, we still made it to the City Council meeting on time!

The structure of the loan proposed in the plan was in three parts. The city of Galesburg would loan $675,000, while a Small Business Administration (SBA)–guaranteed loan of $1.75 million would be facilitated by F&M Bank. While the 80 percent loan was officially backed by the federal government, F&M wanted to have coverage for the remaining 20 percent. Remarkably, the city of Galesburg and its elected officials stepped up to shore up that gap—the third part was a loan guarantee of $355,000 from the city of Galesburg toward this exposed portion on the SBA loan. The council deliberated and unanimously voted 7–0 to approve this package, with the city ultimately loaning us $675,000.

Siva and I offered personal guarantees against this $2.5 million loan. If Intellihot had not been successful and this loan had been called, we would have lost our homes and 401(k)s and would most certainly have to declare bankruptcy—not to mention the garnishment of any future wages for years to come. So, a lot was riding on making this work.

Separately, it just so happened that Galesburg was also home to another recently closed factory: Carhartt clothing factory. Via Cesar Suaez, I met Gretchen Carhartt, the granddaughter of the founder of the company. She has a passion for green and cleantech, and she mentioned an idea she had: taking her empty factory space and turning it into an incubator to promote clean businesses. We talked to her about our company, and she was delighted at the possibility of Intellihot becoming a tenant in one of her buildings.

In 2011, we took our approximately $2.5 million loan, moved into the empty Carhartt factory space, and started our first actual production line to produce a unit for sale—and thus began the next phase of the Intellihot journey.

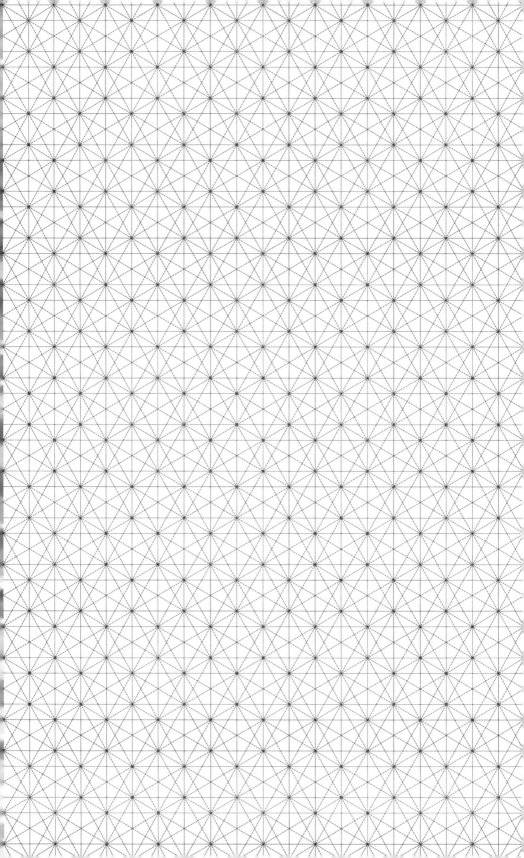

Building a Business, Being a Leader

T he initial days of bringing our water heater into full production were all about developing our supply chain and our production process—while at the same time figuring out how to sell our product in a market that had no idea we existed. It was as though we were a company that had created the best car in years but had no dealership and no mechanics in the world who knew about us. How could we alert the world to our presence and our product?

To begin with, we knew we needed a niche. Early on, we decided we would focus solely on commercial water heating. On the residential side, especially outside of the United States, in parts of Europe, Asia, Australia, and New Zealand, tankless water heaters were already becoming popular. However, there was no commercial tankless water heating product available anywhere in the world. The commercial water heating industry had remained completely stagnant. Nobody had attempted to do tankless water heating—never mind smart water heating—on a large scale, until we came along and began to build tankless water heaters designed for commercial use.

We positioned ourselves in this white space and created a brand. We introduced the brand, built the brand, built our product, built additional products, and reinvented the market. It was not done all at once. We had to invent and reinvent ourselves so many times along the way—and that reinvention continues to this day. But at the core, we stayed true to our dream: sustainability, energy efficiency, health, and safety.

Viewing Problems as an Opportunity in Disguise

Trying to start production and introduce a new product using debt was a tough lesson in economics. With our $2.5 million in loans, we thought, "We are going to produce these units, we are going to sell them, we are going to make money, and we are going to service this loan." In reality, it was nowhere near that simple.

As I have mentioned, we had decided early on that we wanted to develop all the disparate components of our units ourselves, from scratch, and bring them together. This is not an unprecedented model; Apple functions similarly, designing and developing everything themselves, right down to the software. However, Apple uses outside partners, such as Foxconn, to produce the actual units. We wanted to be like Apple, only more so: we wanted to not only design and develop all our own components, but we also wanted to produce the units ourselves at our factory.

Although we would design and develop the components from scratch and assemble the units ourselves, physically making each of the components ourselves would cost millions and millions of dollars. We had the components fabricated and produced by numerous outside vendors, each of whom typically had the capacity to manufacture large quantities and were themselves looking to diversify their business or leverage their extra capacity.

We sourced our components from all over the world: Germany, India, China, Japan, Italy, Mexico, and, of course, the United States. It was truly global—and as we finalized this plan, we quickly realized we needed someone to keep all of those moving pieces running smoothly. So, our very next hire for our new facility was a logistics manager. We found Leah Griffin, a terrific woman who previously worked for Wilson, supplying their sporting goods to Walmart. She wanted to move from Chicago to Galesburg to join us—and to be closer to her future husband, who lived in Galesburg. The stars aligned for all of us!

The initial samples of most components we received were great—high quality and visually appealing. The first couple of units worked fine. Everything seemed ready to scale up to our first big batch of 250 units, and so the first large order of components was placed. However, when we received the components in large, bulk quantities, suddenly, nothing worked. Nearly every component had a problem: the lines wouldn't fit, the valves leaked, the cabinets were not square, and the tolerances were all over. When we put the units together, nothing fit properly, and they failed our end-of-line tests. We were left unable to sell any of the first units we produced—and we still had to repay our loan. We ended up drawing from one loan to repay the other—in other words, servicing our debt with our debt.

We also learned that when it comes to components, less is more. This was not the first time I had learned this lesson; I had encountered it decades earlier, as a child on my father's farm in India. In those days, my uncle would always park the tractor up on a hill and jump-start the tractor by rolling it down the hill. In fact, every time my uncle got a new tractor, the very first thing he did was to remove the battery and throw it away. Finally, I had to ask, "Why did you discard a brand-new battery and why don't you ever key start?"

"The battery," my uncle explained, "is the most unreliable component of the tractor."

This was my first lesson in how important it is to have reliable components and how it is often better to have fewer components—a lesson that was now proving very relevant to my work at Intellihot!

Every day, I would go into the factory, look at all the issues we were having with the components, and say, "Well, this component is no good; what if we eliminated it? That component is no good; what if we eliminated that?" We kept simplifying the design, but it still didn't work reliably, so I kept looking for more components to eliminate.

The unit we initially set out to build as our model #1 was called the pump model, and it was designed to heat and deliver water. The design of that model worked perfectly, but we simply couldn't get functional components at scale—so I started eliminating components. First, I removed the crossover line, then a solenoid valve, then the pump, and then more components, until eventually I boiled the unit down to just a heat engine and the cabinet around the engine—and that became what we called our non-pump model.

When you go from concept to production, you will inevitably have to reinvent—and reinvent quickly. We hadn't planned for this new non-pump model at all, but it ended up becoming the anchor product for Intellihot. It was reliable and heated water very efficiently and was so successful that we still produce it today!

Part of what made this kind of rapid reinvention possible was our foundational idea that engineering and production should be colocated. We were still figuring out how to build everything, we were still figuring out what worked, and we were making changes very, very rapidly. So, it was essential for those two departments to be able to communicate and collaborate in real time. That enabled rapid iterations and innovations.

A Different Way of Working

This way of working was a big learning curve for everyone at the company. We were still a very small operation, with only nine or ten people in total: me, Siva, Mike, our logistics manager, a production manager, someone in charge of testing the units, and a couple of people to assemble the units. With that tiny crew, we had started building twenty or so units at a time to fulfill the small trickle of one- or two-unit orders we had coming in.

Obviously, nobody on the team had built units like this before—this was a brand-new design! Siva, Mike, and I had to figure out the appropriate build sequence ourselves and then train everyone on how to put them together. Then, as the team put the units together, we would inevitably come across an issue and would need to tear those units apart, rebuild, and retest them. We certainly didn't want to send any units out to customers with any kind of issues! This happened over and over, as we continued to scale production and make the units work reliably. On the software side, it was a similar story: the first version of software that worked reliably was version V361; it took 361 versions before one worked reliably without bugs or crashes!

Our team of employees had never been in this kind of environment before, in which they were helping correct problems within the product as they built it for production. "Why are we tearing this up and rebuilding it again?" they would ask, bewildered. "We just built and tested it last week!" To make matters worse, we always seemed to have a shortage of one component or another. This resulted in units or subassemblies being partially built and the rotation of assembly workers to different stations based on what was at hand.

Eventually, I drew out our process on the wall, end to end, for everyone to see. "Our goal is to make the product work and fulfill demand," I explained. "We are going to sell X number of units of various models. We get the various components in various quantities for different models, and each of those components arrives at the factory at a different time. Some of the components have quality issues, so we have to adjust on the fly. Then, when we sell, we sell a slightly different model mix than we thought we would sell. On every node of this entire chain, end to end, there are variations and elements that will shift. We will have moments when we need to rebuild things, reassign people, build subs, but we have to keep working through these variations and keep solving problems."

The environment was highly in flux, changing daily. That made some people uncomfortable, while others enjoyed it! As founder and CEO, I wanted to make sure everyone in the factory felt energized and uplifted. When I worked for Caterpillar, I noticed there were many practices not conducive to employee well-being or morale. Many were good intentioned, but some were downright opaque. I wanted our workplace to be different.

Early on, we didn't use a punch clock or any other kind of clocking system, trusting our employees to show up on time instead. We also kept a lot of food in the factory and placed it where employees would need to walk through the factory to reach it so that everyone would cross paths and interact with one another, creating a sense of shared space and purpose.

To keep everybody's spirits up and energized through the constant reinventions, I bought a giant wind chime and hung it at the entrance to the factory. Every time we would get an order in, I would immediately ring the chime so that the entire factory could hear it and celebrate our success.

First Customers

One of our first customers was Jim Barrack of Barrack's Hospitality Group. We owe Jim a great debt of gratitude: he agreed to try out our very first test unit, giving us our first real-world opportunity to put our water heater into action. Jim ran a commercial food business called "Barracks Cater Inn," which was open from five in the morning until eleven at night. It was a hard-run facility, making it an excellent testing ground.

At Caterpillar, Siva and I had extensive experience testing machinery. Caterpillar machines are built to be as indestructible as possible. For example, a truck had to be able to run 12,000 hours without requiring service. Of course, you can't run a truck in a lab for 12,000 hours, so we tested the trucks on compressed, accelerated cycles and developed methodologies to string together the most damaging cycles a truck might face in the real world. Five hours of testing in the lab were equivalent to approximately two hundred hours in the field.

We followed this same philosophy in testing our water heaters, putting them through rigorous testing—but there was still some risk involved in putting the very first test unit into practice! We must have sufficiently impressed upon Jim that we cared about safety and that we understood how water heaters worked, even though this was a brand-new product. He even magnanimously signed a release of liability, stating if the unit exploded or caught fire and burned down the building, he wouldn't sue us!

Thankfully, the test unit did not blow up or catch fire. However, things didn't go entirely smoothly: the very day after we installed our unit, it popped up a leak error code—and shut down the water for Jim's entire building. The problem, we discovered, was the temperature in the basement. The basement of the building was very cold, and we installed the unit in July, so the fresh air the unit was taking in from outside was quite humid. In the cold basement, that humid air produced condensation inside the cabinet, which tripped the leak sensor within the unit, which then shut off the water to the building. No matter how much testing you do in the lab, there's always something in real life you never thought to account for!

After we adjusted the unit to rectify that mishap, it worked like a charm—and Jim is still one of our customers to this day, using our water heaters in his facilities. Along the way, Jim gave us an even greater gift: he became our reference point for several of our subsequent customers. He recorded energy savings of 44 percent and directly cut CO2 emissions by 44 percent. It was confirmation of our quest to do good for the environment and for our customers.

As we started to produce units for sale, we did not advertise in any big way. The very first unit produced at our Galesburg factory went to a hair salon in Galesburg called Hair Etc. It was owned by Eva Garza, Mayor Sal Garza's wife. To commemorate the start of production in 2011, the mayor paid $2,011 for the unit. I was thankful for their support.

Otherwise, our sales came from word of mouth, from customers like Jim Barrack. Orders for one or two units would trickle in; we powered a hair salon, a couple of restaurants, and a multifamily residential unit. But we didn't really understand how to get our product to market on a larger scale. It was a mystery to us; in fact, I think it's a mystery to the majority of the American public how building products get to market on a large scale. We had a lot of learning ahead of us.

Finally, we landed a large customer—and, true to our unorthodox path, it was a very unconventional customer: the local federal prison in Pekin, Illinois. We had sold them a couple of units, and after trying them out, they found our heaters remarkably suited to their needs.

The prison system has a great need for very good, reliable hot water because the schedule is so highly regimented. Every inmate in the prison takes a shower at exactly the same time, every day—meaning there needs to be enough hot water for every prisoner in the prison at the same time every day. In fact, there is a federal mandate that the water temperature in prisons must be kept at a certain level; a hot shower is one of the few creature comforts inmates get, and they have a legal right to reliably have that comfort.

The prison's water heating system was big, bulky, and antiquated, and every time it would fail—which it did, every so often, completely unexpectedly—it would cause outrage among the prisoners, sometimes even leading to riots. "When the prisoners don't get hot water, they get mad, and they kill each other and try to kill us," the prison administrators told us. "We need an extremely reliable system, because for us, this is literally a life and death problem."

We knew the Intellihot system would be ideal because it is modular, and therefore, it is impossible for it to have a total systemic failure. Sure enough, the prison tested our system, and it proved to be very reliable—and it had a bonus side effect: it cut down energy usage by 43 percent.

The prison had over twenty buildings, each with their own bank of water heaters. They decided to retrofit the entire compound with our system—around two hundred heaters total. Because we were still such a small operation, we fulfilled this order piece by piece—twenty units this month, thirty units the next—until the whole prison was using Intellihot water heaters.

To Sell or Not to Sell

By this time, we had clearly demonstrated that we had a product completely different from anything else on the market—and that we were able to open a factory and build and sell units. All of this caught the attention of one of the country's largest water heating manufacturers—a company known for producing heaters but not known for being innovative or having any products in the market that made much of an impression. They had been made aware of our work by a person we had been trying to hire to lead our sales team. He had gone to the large company and said, "Hey, I've been talking to this super innovative water heating company—you should buy them!"

We started a dialogue with this large company—a dialogue that ended up occupying the bulk of our time for an entire year. Executives from the company would come visit us, and then they would tell us to come visit their plant. We would have meeting after meeting, talking about how to operate our company and how this deal might work. For all of 2011, we accommodated every request from this company, doing whatever they needed to do to get to know us. I did like their CEO and head of manufacturing, both of whom were gregarious and smart individuals. They had navigated many things well and built a large, successful, traditional water heating business.

Despite this, we went round and round, had dinner after dinner, traveled all over, but ultimately, we couldn't get a deal done. This company had been around for over a hundred years, and they were a good company. I understood and respected them. But the more we got to know them, the more we realized there was no alignment with why Intellihot exists and what we were seeking to do.

The intention of the big company was to bolt us onto their existing product line to bring innovation—but also put strong boundaries around us founders. The contract they presented to us essentially stated that we could not do anything other than work on water heating. We could not experiment with motorcycles, or robots, or intelligent machines. In short, they wanted us to think about nothing but water heaters all the time.

What this company didn't understand, and what Siva and I knew both intrinsically and from experience, is you can't imagine new things in a vacuum. Innovation requires input from anywhere and everywhere.

"Everything we have invented is because of our work on motorcycles, and marine engines, and robots," I explained to the CEO. "If we only thought about water heaters, we would never have invented this innovative product. If you put these boundaries on our imaginations, you won't have innovation. Innovation simply won't happen in a vacuum or with tight boundaries."

Unfortunately, the CEO did not see things the way we did—and Siva and I simply could not see ourselves going back to work for a large company, having to overcome all those roadblocks and impediments to our imagination and creativity. It would not help us make our vision come true. We were at a decision point—and, luckily for us, that decision was made for us by the company. The company thought if they walked away from the deal we had more or less agreed to, Intellihot would collapse, and perhaps they would be able to come in and buy us cheaply. This turned out to be a huge miscalculation on their part.

This company had been talking to us for an entire year and discouraging us from launching the product in a big way. They wanted to be the ones to introduce our product to the market as their new, innovative product. So, they discouraged us from attending trade shows or showing the product in any meaningful way. As a result, when the company walked away from the deal in November 2011, our funds were almost down to zero. I had to move at lightning speed. For a moment, it looked like the company's prediction that Intellihot would collapse could come true.

But fate has a funny way of turning things around. As we started looking into trade shows, we learned that the AHR (American Heating and Refrigeration) show, which is one of the largest HVAC (Heating, Ventilation, Air-Conditioning) trade shows in the United States and was just a couple of weeks away, had some cancelations—including a prime spot. Within six weeks of the company pulling the plug on our deal, we managed to book one of the best spots at one of the largest trade shows, showcasing our innovative products.

The big company could not have been more surprised. "We usually plan trade shows for six to eight months," said the company. "How did you pull this off in less than eight weeks?" The answer was simple: adversity had already taught us to invent and reinvent ourselves, and we knew how to pivot and move rapidly. Plus—how hard could it be?

I am still good friends with that CEO, and we talk from time to time. I think he realizes walking away from the deal was probably the wrong decision—but it turned out to be one of the best things that happened for Intellihot.

IoT and the Connected Water Heater

We received a great deal of attention at that first trade show because we were so different, so far ahead of anything else in the industry. We not only had a very different tankless water heater but also the industry's first IoT and app. One could see our units using an app on a Gen-1 iPad. People were curious about our products, the pump and non-pump model, and its IoT and connectivity features. I remember people asking me, "Why would I want to see this on my phone?"

Here is how it came to be: We were designing a remote display. As we thought about it, we asked ourselves, "Why are we designing this separate display, when everyone is carrying a little computer in their pocket?" This was in 2010, and smartphones were just coming into prominence. Why not create a display everybody could have in their pocket so that people could have all the information right there in their hands?

We abandoned the idea for a dedicated remote display and focused on developing an app for the iPhone, and there we were: a company making revolutionary change in an industry that had been the same for a hundred years *and* developing cutting-edge IoT connectivity to go with it.

Of course, people would ask us, "Why would I ever want to look at my water heater on my phone?"

"Well," we would reply, "if something is going wrong with your unit, the app will tell you ahead of time through an alert on your phone. You'll never end up standing in your basement, ankle-deep in water."

Finding Investors

Even with the attention we were receiving at trade shows, we were still in a bit of a rough spot financially. We had burned a good deal of time and money going round and round with the big company. We had been focused internally on building the factory and not focused much externally, because we thought this big company was going to acquire us. Now, we needed capital, so I turned to the path taken by so many start-ups before me: pitching the company to private investors.

At first, we started out very small, but after about a year, it became clear we needed larger investments in order to scale up. Intellihot had a couple of challenges to overcome when it came to attracting major investors. First was our location: we were in central Illinois, an unlikely place for an innovative start-up compared to Silicon Valley, which is full of start-up companies. We were literally in the middle of nowhere, with no access to the kind of venture capitalists Silicon Valley companies or companies in Boston, Chicago, or New York have. If we were located in California, I know we could easily have attracted investments of $10–20 million.

Second, we were not in what you would call a "sexy" space. We weren't inventing an exciting, sexy new software or app; we were inventing a new water heater, and water heaters are decidedly unsexy. Yes, it was cleantech, but that wasn't the hot topic it is today. People were aware of the earth's existential energy crisis, but it wasn't popularly considered a pressing or an immediate issue en masse. Investors were much more interested in software and apps, something that could be scaled fast and provided quick returns and didn't require real iron and a real factory. These factors put us at a disadvantage.

The interest that existed in cleantech investment had also taken a blow in recent years, making 2012 to 2015 a particularly difficult time to raise money for cleantech start-ups. After the 2009 stock market bubble burst, the dot-com tech industry was down and out. As a result, many Silicon Valley companies rapidly invested in cleantech, without really understanding what cleantech is and, in particular, how hardware intensive it is. Cleantech is not a software industry; it requires the development of physical hardware, which takes longer to get right and doesn't scale as quickly. Cleantech can't follow the same model or timeline as software or app development. Because of this, many investors in cleantech became discouraged after two or three years, and there was a large pullback from investment in cleantech—just as I was trying to raise money for my company.

All of this was further exacerbated by high-profile bankruptcies like that of the solar company Solyndra. The federal government had invested a good deal of money in cleantech, and despite pitfalls like Solyndra, they actually got a positive return on their money. But this did not translate into private investors returning to cleantech start-ups.

Despite these disadvantages, we also had some qualities that were very attractive to investors. First, Midwest companies are generally a bargain and get lower valuations than Silicon Valley companies, meaning investors can invest at a lower level for equal returns—one benefit to being in central Illinois! Second, we had a verified, tested product that was proven to work, and work well and reliably, along with real customers and our own factory space. Third, the founders—

Siva and I—personally signed off on our product and company. We were personally liable for our $2.5 million in loans; the bank could take our homes, our savings, and our 401(k), if we defaulted. This showed investors the incredibly high level of trust and belief we had in our product and company—and it made us look like a pretty solid, low-risk deal!

For the first time, we started to get some real investor money coming in. It wasn't much, and it came in bits and pieces—$250,000 here, $50,000 there—but it was something.

Lessons from the Investor Search

The process of looking for investors was also a great learning opportunity. As we saw other entrepreneurs and CEOs pitch their companies, we started to develop a sense of how to pitch our own company. We also developed a sense for which companies and teams would succeed and which would not, based on their pitches. We saw many company leaders give terrific pitches, sounding incredibly smooth and polished. Siva called them "the CEOs with the slicked-back hair."

Although we were definitely not "CEOs with slicked-back hair," we learned a great deal from their pitches. We saw one pitch in which the CEO talked about a medical product having coast-to-coast testing—but, in actuality, they had one customer in New York, one customer in Chicago, and a potential third customer in California they had not even reached out to yet. However, "coast-to-coast testing" sounds much better than "We have three customers!"

We quickly observed that pitching is all about presentation. While we were always completely honest, we learned how to present ourselves and our company in the best light, in a way that would appeal to an investor.

When it comes down to it, investors have only one goal: to make a return on their investment. This is not a bad thing; they want the company they are investing in to succeed so that it provides them with good returns. Therefore, it is important to be clear on the return on investment (ROI) and to have no illusions about why the investors are there and what they bring to the table. That being said, I would also often look for investors who were sympathetic to the cause of energy efficiency and the environment. Some had funds dedicated to those themes, which could further catalyze them to invest in Intellihot.

But the most important lesson I learned when it came to investors was to trust my gut instinct. As we started to pitch to angel investors, they would often tell us we needed to hire a new CEO, a new CFO, a marketing firm, this guy, that guy. "If you don't," the investors would warn, "you will definitely fail." Their view was the best person to take a company from zero to the first hundred is not the inventor; it must be a seasoned CEO.

This made no sense to us. Nobody knows more about their technology than an entrepreneur. Nobody knows more about their market than an entrepreneur who tried to solve problems in that market. Nobody is going to put in more time and effort than an entrepreneur. Investors can't see this, because they tend to look at a macro level. They're looking to see if the wind is blowing in the right direction, if you are committed, if you are frugal, if you are efficient, if your team can really pull it off. They may have a general sense of the industry and overarching market trends, but they often don't know the details of how the product or technology works or what the market space is like for that exact product.

A seasoned CEO may have more experience running a company, but the passion an entrepreneur brings to their company simply cannot be hired out. I am sure there is a point, when a company becomes big enough, where it can benefit from the expertise of a CEO who has more experience running a big company—but I firmly believe that cannot be at the genesis of the company or during its initial growth.

In fact, I have seen this practice fail, more than once. We saw several other promising technology companies that were part of the Peoria Next Innovation Center work with angel groups or investors and follow their instructions to bring on new leadership, only to fail very quickly. I saw a medical company go bankrupt because the founder brought in an outside CEO. I saw an industrial tech company suffer the same fate for the same reason. So, Siva and I shied away from taking any money from these angel investors. We just kept plowing forward with whatever cash we had.

However, we did still need funding—and in one notable case, I made the mistake of ignoring my gut feeling. I had a meeting with an investor recommended by an earlier investor. When I talked with this new investor, it did not pass my gut check. I did not have a good feeling about him. But my first investor was insistent, because this new investor would come in now with some money and then also do a follow-on larger investment. So, I ignored my gut and accepted the investment.

As I feared, the partnership did not go well. To begin with, the investor asked for excessive detail on everything—something I have always been wary of. After all, there was still so much we didn't know ourselves, so much we were developing and working on as we grew. That's the nature of developing a brand-new product! The investor would then say things needed to be changed this way or that way— none of which was useful in making the company successful.

Second, the investor had pledged to invest a greater amount of money if we reached a certain revenue level the next year. The investor clearly did not expect us to reach the level he set—but we far exceeded his expectations, growing the company fivefold that year. "Look," we said when we had our annual reports in hand, "we hit your target. Now we need that next investment from you to scale up."

"Well," the investor said, already backpedaling, "what I really meant was net sales, and if we look at the net sales number, you were off by $25,000." He backpedaled until he walked all the way back on his commitment. A couple of years later, this same investor created more trouble, when he deliberately misinterpreted some documents, arguing over one or two words—with which none of our other investors took issue—in an effort to grab more shares and company ownership.

Behavior like this sours what an entrepreneur is trying to do— and impedes success, for both the company and the investor. All of this could have been avoided if I had listened to my gut at that very first meeting.

Despite this road bump, we did continue to find good investors who worked well with us and believed in our company, which allowed us to take Intellihot to the next level.

The iQ Series—World's First Modular Water Heater

As we started to grow the company with money from investors, we focused on improving our product, acquiring more customers, and building our brand name. The first product we released was very compact, about the size of a shoebox. Rather than building larger and larger water heating units for larger buildings, we built a modular box to which we could add more modules, stringing them together and designing them to intelligently talk to one another in order to behave as a unified group.

I was fascinated by robots made by the company iRobot, which was started by two professors at MIT and became famous for creating Roomba. One of the robots iRobot created was called PackBot, which functions by having a group of robots come together to do intelligent tasks. The amount of software in each robot is surprisingly small—only 16 or 20 kilobytes. There aren't gobs and gobs of code packed into each robot, but together, they can collectively achieve a high level of intelligence and perform complex tasks.

This concept really struck a chord with me, and we applied it to our water heaters: instead of building large singular behemoths, we build smaller modules which are intelligent and robust. As individuals, they have limited capability—but as a group, they are able to handle very large buildings with higher and more complex loads. Our large machines were made out of smaller building blocks of heat exchangers and controls—much like building a giant structure out of many small Lego pieces.

At trade shows, people were confused by how small the unit was, and I would explain that the unit was a commercial unit built to be modular: if you put two of these units together, you could power a twenty-unit condo building.

"This looks just like other residential units I've seen," people would say. "Why is yours more expensive? What makes it different?"

"Well," I would explain, "you can't power a commercial building with a residential unit. You would need three or four times as many units—and they would likely fail within six months. Our units can power commercial buildings, reliably."

After I explained this two or three times, I realized we had a problem with perception: when people looked at our unit, they saw a unit that looked like a residential unit. With this input, I came back from a trade show and sketched out an updated design. Rather than having the engines in separate boxes that could be combined modularly, we put several engines inside a single, bigger box, which had singular external connections. Now, it looked like what people thought a commercial unit should look like, with larger water and gas connections.

Overnight, people understood the concept: that the heater was modular and that a single box could power a large hotel or hospital. But if one of the multiple engines inside the cabinet malfunctioned, the other engines inside that single box could take over so that the system wouldn't fail. That's what made it so reliable.

This experience showed me the importance of perception—and of listening to potential customers. When you bring a product to market, you are responsible for everything your customer sees, hears, or feels, meaning you need to actively and deliberately drive every single customer point and interaction. This includes the obvious ones, such as user interfaces, websites, and apps, but also extends to product manuals, labels, packaging, customer interactions, and customer training. I learned how important it is to stay in direct contact with customers and take note of what they see and feel. Based on my interactions with potential customers, I was able to make adjustments to change the perception of the product—just another way Intellihot reinvented itself!

We launched this model in 2014, and it became an industry first, setting the company on a trajectory of growth. From there, we continued to build up the company and the brand name, expanding our product line based on that model of modularity—until we reached the range of products we offer today. (If you want to learn more about the unique mechanics of the Intellihot water heater, see the Appendix at the end of this book!)

CHAPTER EIGHT

Air and Water: The Nexus of Energy, Health, and Safety

Throughout this growth, we had to continue to push the long-stagnant water heating industry to accept this groundbreaking new way of doing things. But I was no stranger to an uphill battle. A relentless, never-give-up attitude has always been part of my character—something I learned about myself when I went toe-to-toe with a simple salt block.

The Salt Block

One day, while my wife and I were walking out of Home Depot, I spotted a stack of the large rectangular blocks of Morton Salt used for water softeners in a corner. The blocks had a dimple on top, were about eight inches square and twenty-four inches tall and weighed fifty pounds. I casually remarked to my wife that she should never pick one of those things, because they are nothing but trouble! My wife immediately asked me, "Why are they trouble?" Even though we had been married for a couple of years, she didn't know this story. So, I proceeded to tell her the story of the salt block.

Many years before my wife and I were married, I happened to see one of these things at Home Depot. I usually bought the big bags of salt pellets for my water softener, but when I saw this singular solid block of salt, I thought it must be something better. The idea of a single large block versus pellets was intriguing, and I imagined it might even be easier to load into my machine. So, I decided to try it.

Much to my dismay, when I took the salt block to my house and into my basement, it did not fit into the opening of my water softener. I presumed I could just break it into pieces and load the small pieces into my water softener. So, I went back upstairs to my garage and brought back my steel pickax, which had a pointed end. Despite throwing increasingly harder and harder blows on the salt block, it barely made a dent. The salt block looked pristine. I was immediately perplexed at the strength of this salt block. I had fully expected it to crack on impact, but it did not.

"What if I drilled a series of holes into it?" I thought. "Then I could drop it from a height causing it to break." I went upstairs and got my electric drill and a set of drill bits. I set out to drill a few holes in close succession. Again, much to my shock, the steel bits did not quite drill through and, in fact, appeared to be dull. The bits I had used were old, so I went back upstairs again to get some newer drill bits. The results were the same: I was unable to drill into the salt block with any degree of success. At this point, I figured it might be easier to saw the whole thing into two. So, up I went again to my garage to bring down my reciprocating electric saw, which had a six-inch blade. I went at the salt block with full force at various speeds. Again, much to my dismay, the electric saw made no dent. I went upstairs yet another time to get some brand-new blades. I ended up with the same result.

By now, images of the Titanic and the iceberg were conjuring in my head. Could a block of salt be more powerful and stronger than steel? I knew I had to keep going and was fairly confident that I could get it to break. Of course, I had considered dissolving the whole thing in the bucket of water—but that would be too easy.

So, up I went again back to my garage, and this time I decided to use the hardest substance known to man: diamond. I grabbed a set of diamond-tipped cut-off wheels that I had purchased for my Dremel set and went down to the basement. I managed to cut a few grooves in the salt block, but it still appeared pristine. It was nowhere close to breaking into two large halves, let alone smaller pieces.

So I considered another approach.

What if I swung my large sledgehammer at the salt block? I carried the salt block back to my garage and placed it on the concrete floor. I grabbed my thirty-pound hammer and swung at it with full force. I tried every face of the salt block, both the small and the large faces. There was very little damage. The salt blocks simply stared back at me, unaffected and strong. I postulated that maybe the faces might be dissipating the load, and what I needed was to find a weak point on the salt block. "The corners ought to be a weak spot," I thought. I imagined that if imparted a quick blow, otherwise known as an impact load, on that weak point, it would definitely crack. Impact forces multiply the effect multiple fold versus a steady state force applied to the point.

So, I carried the fifty-pound block of salt to my driveway and dropped it on each of the corners—with no effect. Then, I increased the height from which I was dropping the salt block. I went from three feet to nearly six feet. I even fine-tuned the drop at various angles on each of the corners for maximum impact. There was very little damage. Needless to say, by now I was physically exhausted from throwing the salt block and lugging it all over the place doing these various experiments. More importantly, I was in awe at its strength and resilience. I had spent a good part of my day doing this and certainly destroyed numerous tools worth far more than the $9.50 the salt block cost me.

I decided it was something to keep. Perhaps I could come up with a better way one day. So, the salt block went back to my basement and sat next to my water softener for many, many years. I had developed a sense of respect for the humble salt block. It served as a reminder for how something so innocuous can be so tough.

Upon hearing this story, my wife burst out laughing and was beaming with admiration. She said she now understood a part of me, one that was relentless, with a never-give-up attitude. I personally had never thought of it that way, but certainly having the characteristic of not giving up is very useful when you try to invent something and change a long-stagnant status quo.

The Reality Distortion Field

When it came to shifting the water heating industry over to the Intellihot approach to water heating, we began with data and facts—which turned out to be much more effective than any of the techniques I had used against the salt block!

Conventional wisdom states that in large buildings, such as hotels, everyone is going to take a shower in the morning. Therefore, you need a heating system large enough to provide enough hot water for everyone in the building to shower at the same time. That is the only way, the conventional wisdom goes, to ensure the building can fulfill the demand for hot water.

This conventional thinking immediately struck me as improbable, but I couldn't put my finger on why. Then, as I was driving to work one day, I stopped at a railroad crossing and started thinking about how the human brain works. Many accidents happen at railroad crossings because when a driver sees a huge, bulky engine coming toward them, the engine looks like it is going slower than its actual speed. The driver, therefore, thinks they have enough time to make it across the tracks before the engine arrives. They don't realize how quickly the train is approaching.

The brain plays this same trick when people look at large buildings. They make an assumption about how that building will operate—an assumption such as, "Everybody is going to take a shower at the same time." However, if you really think about it, it is apparent that buildings don't operate with 100 percent of the fixtures open at any given time.

In 1940, an engineer named Roy B. Hunter developed and published an estimate for the National Bureau of Standards on the quantity of hot water a building of a given size needs at any given time. This same estimate has been used ever since. The only adjustments engineers have made is to increase the estimate. Engineers are trained to overestimate to make sure equipment doesn't fail and to design equipment to handle too much rather than too little. Then, when engineers hand those designs over to contractors, the contractors say, "Well, I want to make sure this building will never run out of hot water, so I'm going to make it just a little bit bigger, just in case." By the time the water heater actually ends up in the building, it is likely three or four times larger than it needs to be, which is how buildings end up with enormous systems sitting around wasting energy by heating much more water than is necessary.

I wanted to find data to dispel this decades-old theory. When we first released our product, because of this old estimate, people were fearful it wouldn't have enough capacity. After all, our competition was saying, "You need ten of our units," while we were telling customers they only needed two of ours. More often than not, our customers would say, "Well, let's install four units just to be safe."

As we developed our product, we had some discussion about whether the units should be used in conjunction with water storage tanks. We decided to collect some data to see whether storage tanks were necessary and asked some large buildings if we could monitor their water usage with flow meters—something nobody had done before.

The results were eye-opening. We found that, at its peak, a hundred-room hotel would have only 12 percent of its rooms using the shower at the same time. We measured numerous hotels—hotels with fifty rooms, with one hundred rooms, hotels that were fully occupied, hotels that had laundromats. Then, we expanded our monitoring to measure multifamily units and restaurants. We quickly found that we had no need for water storage and decided to make that a central premise of our product.

All of this data became core to developing the size of our product and our estimates of how many units a building would need. The understanding of how to estimate the amount of hot water needed at any given time is also known as the theory of diversity. The theory of diversity is really a theory of probability regarding how many fixtures will get turned on simultaneously. It is intuitive that if you design a water heating system to satisfy 100 percent of all fixtures being turned on, it will result not only in a massive system (both physically and financially) but also in enormous energy waste.

In addition, the type of application (hotels versus restaurants versus stadiums) has a discernible difference in the diversity or probability of fixture use. Especially in new construction, there are insufficient data points to account for all the factors that impact hot water demand. Defined design criteria are laid out in the ASHRAE (American Society of Heating, Refrigerating and Air-Conditioning Engineers) guide and the Uniform Plumbing Code (UPC). Both criteria focus on the use of probability theory with a safety factor to compensate for unknown variables.

Thus, the next logical question is, "How does one design a system that is large enough to meet any and all anticipated loads yet small enough to be economical and energy efficient?" Here, rather than giving in to fear of the unknown and the resulting oversizing, we propose a data and facts approach, one that is augmented by field measurements and a review of existing sizing methodologies proposed by ASPE (American Society of Plumbing Engineers), ASHRAE, and prevailing sizing tools from other manufacturers.

Utilizing field measurements from various applications across the United States and Canada, as well as accounting for all the above factors, Intellihot developed a sizing calculator. We believe this calculator presents the user the optimum sizing, one that drives capacity shortage risk to a minimum, while maximizing energy and space savings.

With this tool, we introduced the Intellihot guarantee: use our sizing and methodology, and if we ever fall short, we will give you additional pieces of equipment at no cost. We have had this guarantee since 2015—and we have not had to deliver on it even once. What's more, our guarantee pushed others in the industry to provide their own guarantees for large buildings!

The Intellihot Value Proposition

This guarantee of reliability is central to Intellihot's value proposition, especially in the water heating space. Companies are built to serve customers, so reliability is of the utmost importance. In fact, reliability is so important; it is the main reason the water heating industry hasn't changed in a hundred years. For a hundred years, water heaters have worked just fine, and nobody wants to risk changing something that works—even if it fails after a certain period of time.

In this way, the water heater industry is like the airplane industry. Because of my love of airplanes, I was at one point thinking about trying to build my own plane. I started looking into kits and discovered that one of them used the exact same engine that was designed in the 1930s and had not changed at all since then. Why? Because this engine worked reliably—for 8,060 hours. It was completely dependable for that amount of time, and then it would be time for an overhaul.

This is how the water heater industry feels about the traditional water heater model: it works reliably for a certain amount of time, and when it fails, you replace it with another tank that works reliably for that amount of time. This system works at the expense of efficiency, at the expense of space, and at the expense of providing healthy water—but otherwise is generally reliable. Coming from outside the industry, we took a whole new approach to reliability, which led us to create our modular unit.

But reliability is only one part of any company's value proposition. The second is cost: how much does it cost up front and how much does it cost to operate? This part of the value proposition is also what sets Intellihot apart. With our system, you spend less up front and less in operating costs, *and* you have a more reliable system. Moreover, because you are not storing large amounts of stagnant water, the risk of *Legionella* bacteria is mitigated. So, you are also providing your customers with good, clean, efficient, and healthy water.

This value proposition is not hypothetical; it is something we have seen in practice with our customers. We powered a four-hundred-room Hilton hotel outside of the San Francisco airport, and with our system, they cut their gas bill by over 70 percent. They had been using around twenty million BTUs to heat their water. We told them that with our system, they would only need five million BTUs. The engineer designing the building was worried about using something that small, so he doubled it, and our sales rep who was selling the units didn't resist—after all, he was selling twice as many units! So, we ended up putting in eleven million BTUs worth of units. All of the eleven million BTUs never get used, but because the unit heats water on demand, they don't waste any energy even though they are oversized. The hotel's monthly gas bills went from around $20,000 to $5,000, and we eliminated about 8,000 gallons of stored water. Now, the system only stores ten gallons of water. They have a more reliable system, they save on operating costs, and they saved on their initial investment.

Similarly, we also powered a sixty-two-story luxury condo building in Chicago—one of the tallest residential buildings in the city. When this building was built, it was designed to be energy-efficient and LEED certified. The building had gigantic two-thousand-gallon storage water heaters on the sixty-second floor. When we replaced them with our system, it cut the cost of water heating by almost 30 percent. Moreover, people living in the building reported that the water temperature was more consistent and that the water tasted better!

Even in an industry as stodgy and unchanging as the water heating industry, a value proposition this significant is hard to ignore. When people hear our pitch, they start to pay attention: we cost half as much, we cost less to own, we occupy less space, we save you money on energy costs, we are more reliable, and we won't get people sick from unsafe water. These attributes are our core operating theme, and we continue to build the company today based on this value proposition. And all of these attributes intersect in the nexus of three elements: energy, health, and safety. And these three elements begin with the air we breathe and the water we drink and clean with.

Air and Water Inside

When I first moved to the United States, I was baffled by the fact that every house, every building, sometimes every room was required to have a carbon monoxide detector. I was also baffled when I noticed how prevalent air filters were—and how frequently they need to be changed. Was the air we were all breathing every day inside our homes and workplaces really that bad?

As it turns out, the air we breathe indoors is actually more polluted than the air we breathe outdoors. I discovered just how bad it was when we were designing our water heating units and considering which materials we wanted to use. As part of our research, we analyzed indoor air—and realized we had to pick certain materials for our units because the air inside was so bad that it would actually damage the materials in the unit, negatively impacting the water heater.

Why is our indoor air so bad? Our homes are not well venti-lated, and we are constantly releasing harmful aerosols like hairspray indoors. But that's not the only reason our indoor air quality is so poor: it also comes down to how most buildings and homes are heated in this country.

Anytime you move air, you want to do it in such a way that you don't introduce additional particles and allergens into it. The current furnaces people commonly use, especially natural gas-fired furnaces, heat air essentially the same way a skillet on a stove heats vegetables: with the flame on one side heating up the air on the other, just as flame on one side of a skillet heats up vegetables on the other side of the skillet. If you leave a hot skillet on the stove with nothing in it, you can actually smell the burnt metal. It doesn't smell good! But that is exactly how air is heated with the gas furnaces predominantly in use in the United States.

Anecdotally, I have found a whole range of issues with this type of design. It is a huge carbon monoxide risk because the metal can get punctured—which is predominantly why the US code requires carbon monoxide sensors inside buildings. This kind of furnace also causes the air to become incredibly dry, which ends up carburizing the dust in your house, turning it into finer particles, and degrading air quality.

As you can see, this method of heating air actually introduces more problems than it solves. It makes your air dry, so you need a humidifier. It creates more dust, so you need better filters. Just about the only advantage these heaters have is they are simple to manufacture and are super cheap. Apart from that, they are bad for health and bad for the environment. And yet, the companies that manufacture these furnaces don't address these issues of health, safety, and energy. They pass the responsibility off to the companies that make air filters and humidifiers and carbon monoxide detectors.

If we approached heating water the way other companies do—the way other industries approach their products—we wouldn't care about the quality of the water. We wouldn't care if there were bacteria in the water. We would consider that to be a job for someone else, such as a water purification company, just as the furnace industry considers air purity to be someone else's problem.

At Intellihot, instead of passing the responsibility off to someone else, we developed the on-demand and tankless method of heating water which does not foster bacteria in the water. We created systems that keep water circulating and learn your behavior, so water doesn't need to be stored and left stagnant for bacteria to grow.

We consider this part of our responsibility. It is the responsibility of the water utility company to bring good, clean water up to the point of entry to a house or building. Once the water is inside the building, it is the responsibility of the homeowner or building owner. We often encounter situations in commercial buildings where the water is fine up to the point of entry, but the quality decreases once it is in the building. Our units help the homeowner or building owner keep the water in the building safe and healthy.

COVID unexpectedly revealed many of the shortcomings of water circulation in buildings. When the pandemic shut everything down, buildings were suddenly unoccupied for long periods of time. Without the water in the building being consumed, it sat stagnant in the pipes. Cities and municipalities treat water with additives like chloramines to keep it safe, but those additives are only active for a certain amount of time. After that time, they expire and break down, leaving the water unprotected and open to contamination.

Obviously, COVID was an extreme situation—there is no safety standard in place for constructing a hundred-story building that is normally occupied but suddenly becomes vacant for months.

However, it highlighted the fact that once water enters a building, it is up to the building owner to keep the water safe, even in normal circumstances. A study on extended stagnation in buildings caused by COVID shutdowns, published in *AWWA Water Science* in June 2020, found:

> Normal building operation can often result in stagnation (e.g., offices over the weekend, unused hotel rooms), [stagnation] reactions ... continuously occur at highly variable rates. Green buildings may be especially impacted by stagnation as they are designed for lower water use without substantially changing plumbing design. To manage the water quality issues that occur even with normal use, some buildings (e.g., health care) are required to have building water management plans (BWMPs). However, in a small survey, nearly 60 percent of commercial building respondents ($n = 29$) had not heard of building water management plans.[4]

In a large building, water could potentially be circulating through the pipes for a good twenty hours before anyone uses it. It is up to the building owner to ensure that water is safe for consumption—and if a building uses our units, we also consider it our responsibility to make sure the water we heat in our units is not making it unsafe. That is why we are releasing a new technology—a unit called the Legionator—that has an integrated ozone disinfection system that treats the water at the point of use, such as a hand wash faucet, keeping it safe.

We don't have to do this. Traditionally, we would consider bacteria in the water to be someone else's problem; our only job is to make the water hot. But we think about the system holistically—so even though Intellihot doesn't have a long history of water treatment, we have done our homework and figured out how to treat water with ozone to make it safe.

Holistic Approach to Our Built Environment

So many issues around energy, health, and safety could be easily solved if people just looked at them holistically. The problem stems from us all too often looking at these elements in isolation. Many companies are monochrome: they only do one thing.

The HVAC industries have all evolved separately from one another. The companies making water heaters typically will not make something that heats or cools your house. The companies making heating and cooling devices for your house will not make water heaters. This siloed thinking ends up creating so much unnecessary waste, is bad for our planet, and is unhealthy for us.

For example, air-conditioners absorb good heat from your house and throw the heat outside. Meanwhile, another device, such as a water heater, consumes energy to push heat into your house. Water heating and air-conditioning are both multibillion-dollar industries that exist entirely separate from each other. Instead, these systems could exist symbiotically: if you are air-conditioning a house, you can take that heat and put it into heating water. They don't need to be separate devices.

There are numerous devices in the HVAC and power-generation spaces that are piecemealed together for commercial and residential buildings. But while there is some minimal level of integration, it is not enough to make significant inroads into energy savings, environmental benefits, or our health.

Holistic systems are a big opportunity. There are many terrific engineers working at these companies who could certainly come up with holistic systems, but it needs an innovative type of thinking—an expansion and integration of the thought process. People are siloed into thinking only about their particular area. In order to make the world better, we can't have siloed thinking. Engineers, designers, companies, and entire industries need to start talking to one another and working together. It might be uncomfortable, but they all need to get uncomfortable and stretch themselves outside of their comfort zone. Vision and support from top management are also needed.

The systems of the future cannot and must not work in isolation. When we heat or cool our homes, we cannot absolve ourselves from the effects outside our homes. Companies that produce furnaces can no longer absolve themselves of the damage to air quality they cause. No longer can they think, "My job is only to heat air. It's someone else's job to add humidity, and someone else's job to remove the dust I created."

All of these systems could be collapsed into one—and several should be eliminated. If the furnaces were designed in a different way where they did not pose the risk of releasing carbon monoxide, our homes would be much safer. If they were designed so that they didn't dry out the air or produce more dust, our homes wouldn't need humidifiers or such strong air filters.

Certainly, advanced control systems and building management systems can help. However, they will have limited impact unless the root cause of inefficient energy transformation is solved.

The WELL Building Standard and Regulation

When we design homes and buildings, we have to take into account the health and wellness of the occupants. To address this, the US Green Building Council has certified the WELL Building Standard, "a performance-based system for measuring, certifying, and monitoring features of the built environment that impact human health and well-being, through air, water, nourishment, light, fitness, comfort, and mind."[5] At Intellihot, we are big supporters of the WELL Building Standard because we have already subscribed to the philosophy that energy, health, safety, and well-being are interconnected.

Standards like this are an important part of creating healthier, safer, and more energy-efficient appliances. Industries like water heating are largely made up of legacy companies that have huge lobbying power and usually lobby to keep things as they are. For this reason, standards often lag behind technology. Commercial water heating standards were just updated after nearly a decade with no big changes. They're not going from good to great; they're going from bad to somewhat better—and they aren't being implemented until 2026. Meanwhile, we are saying, "Why can't they be implemented in 2024? We have the technology to do it now, so let's do it now."

I am a strong believer in the free market and letting economics drive innovation, but there are upsides and downsides. The upside is when companies win on the value proposition—when they are bringing customers what the customer wants—it drives innovation. But companies can also charge more for devices people look at every day. This encourages companies to make those innovations in order to make a greater profit.

People are more likely to pay more for an efficient, better-looking, better-working refrigerator or cooking range than they are for a furnace or water heater they never see—even though a furnace and water heater have such a direct impact on their health and well-being. Appliances like furnaces, water heaters, and HVAC devices, which sit unseen and yet have a large impact on health, safety, and energy, often require more regulation to drive improvement.

The fact is, anytime you design something that transforms energy, it is going to impact people's health and safety. I believe if companies and engineers kept this in the front of their mind as they were designing a product, rather than simply thinking about producing a device, they would create a safer and healthier product. If the systems of the future are more integrated, it will make them safer, healthier, and more efficient.

The Next Level of Energy

At Intellihot, we have a penchant for introducing devices that are not just slightly better but completely redefine what the device is. We are not just going to introduce a better water heater; we will introduce something that stores energy, heats water, and provides electrical power all at the same time.

We do this because we believe there is a nexus where energy, health, and safety all come together. One of our visions is to create technology that combines water heating, space heating and cooling, and electric power.

In 2023, we launched Electron, the world's first tankless heat pump water heater for commercial and industrial sectors. This device features a grid-connected thermal battery that can store heat when the cost of power is low and utilize it later to heat water. A thermal battery is exactly like an electrical battery, except it stores heat instead of electricity. We absorb heat from the air and store it in the thermal battery. Cold water flows in and comes out as hot water. We can charge the thermal battery using electricity from the grid, but we can also charge the thermal battery using solar power. On top of the unit, we'll have a solar panel to absorb heat from the sun and directly store it in the thermal battery. This stored energy then heats potable water, on demand.

The unit can operate in several modes:

- **standard mode**, in which the heat pump moves energy from external air to the thermal battery and that energy is used to heat water on demand

- **self-learning mode**, in which the thermal battery is charged at optimum times based on anticipatory loads

- **low-cost mode**, in which the thermal battery is energized utilizing lower-cost, grid-connected power

The unit automatically operates in these modes. During most of the day, it will extract heat from the air and store it in the thermal battery—that is, standard mode. The unit is also always checking whether it needs to buy additional electricity from the grid, depending on the time of day. Late at night, the electricity price is cheaper because people aren't using as much electricity, so the unit can buy electricity when it's cheaper and then store it in the battery as heat. During a summer day, when the sun is overhead, electricity is more expensive because everyone is running their AC—but there is an abundance of solar energy. The unit automatically decides where it should source its energy from to result in the lowest operating cost and impact on the environment.

The Electron is a modular unit that will start out as a singular device that can heat and deliver water, but future versions will take care of all energy needs in your home. This unit will have zero emissions, except for whatever emissions are produced by the grid. If the grid is powered by natural gas, there will be emissions associated with that electricity. If the grid is powered by solar or hydro, there will be zero emissions, even when the unit is using electricity from the grid. It will be available in 2026.

Air, water, energy, machine intelligence, efficiency, health, safety, and well-being are all interconnected. For me, Intellihot represents tying all of these pieces together, which is, to me, the epitome of what we should all be working toward.

CHAPTER NINE

The Climate Is in Crisis—and We Are All the Solution

Air and Water Outside

Anytime you transform energy, it has an impact not just on the environment within the building but the outside environment as well—the very planet itself. And yet, in general, the building technology lags far behind many other industries when it comes to emissions—even though buildings are the largest consumer of energy in the world,[6] accounting for a third of global energy consumption and one-quarter of CO_2 emissions.

When I worked at Caterpillar, one of my roles involved making sure the machines were compliant with the Tier 4 diesel engine standards set by the EPA, so I was always attuned to what was coming out of the engines of our units. I am therefore very aware that today, most of us are driving cars that utilize only 20–22 percent of the energy that goes into the car. But right now, we are at an interesting point in human history where many car manufacturers are pledging not to make any more gasoline-powered cars after 2035. Of course, gasoline-powered cars will still be on the road a long time after that, since cars have a long life span, but they will slowly fall out of use.

This is an incredibly important development, because the internal combustion engine is one of the worst things we ever invented. They are terribly inefficient: the amount of energy we get from one gallon of gasoline is equal to 450 hours of human labor.[7] Imagine going to an individual and saying, "Work for me for 450 hours, and I'll pay you the cost of a gallon of gasoline: $4.50."

Who would do that work? Nobody. We have taken an energy source that is highly compressed and dense—energy that has accumulated over billions of years thanks to organic matter converted through sunlight and condensed over thousands of years to form this compact petroleum—and we burn it and release it in an instant.

That tremendous amount of energy is what has powered us since the Industrial Revolution. From growing crops to packaging products, it all comes from petroleum. The problem is enormous and widespread. We cannot even farm without petroleum. Bullock carts or oxen-powered plows cannot feed so many people. If you removed petroleum from farming today, we'd all starve, because we would not be able to farm. But at least in automobiles, we are starting to develop a path to leave petroleum behind.

The automotive industry, and just about every other industry apart from the building trades, has found ways to build things better, more efficiently, and cheaper. The building trades industry, on the contrary, has not improved much in construction efficiency or safety, and yet the price has skyrocketed. Today, $300,000 allows you build a much smaller home than it did ten years ago. If the automotive industry followed the same pattern, a Honda Civic, which cost $13,000–15,000 twenty years ago, would cost $45,000–50,000 today. I don't know about you, but for that much money, I'd want to buy a fancy sports car, not a Honda Civic—even though it is a car I like!

While there is much discussion in the auto industry about moving toward lower or zero emissions, that conversation was almost entirely absent from the water heating industry. As I entered into the world of water heating, this absence struck a chord. The building trades contribute a great bulk of the world's emissions and are generally very inefficient. Even "eco-friendly" Energy Star–rated tank water heaters operate at only 60–70 percent efficiency in real-world conditions because of the transient nature of hot water usage, standby losses, and efficiency loss because of scaling and high return water temperatures. The steps that home appliances have taken are still so far behind other industries.

When we founded Intellihot, I knew overall home energy efficiency was a high priority for me—but we needed to narrow down on how we wanted to attack the issue. Initially, we were thinking of creating a device that would produce hot water, heat and cool a home at the same time, and potentially produce electric power as well. However, it quickly became clear that we could, and should, only concentrate on one thing to start out, so we narrowed our focus down to water heating, specifically commercial and industrial sectors.

When we started our journey to create a new way to do commercial water heating, we knew we wanted to bring into the built environment heating and cooling aspects that were light-years ahead—accounting for energy efficiency as well. When we built our first devices, we were very attuned to making sure we put the least amount of NOx into the air possible. In order to truly have no emissions, our devices would need to be entirely electric. But over ten years ago, when I first looked at powering devices with electricity, I quickly came to the conclusion that to do large-scale commercial water heating, we could not power our units solely with electricity with the current grid infrastructure. However, this is starting to change.

In order to heat the amount of water consumed by these buildings, you need a lot of power. There are two ways to get power: first, you can get all that energy at once and store it in a large tank or battery and use it when needed. The other way is to draw power when you need it to heat on demand. It turns out, water heating is so energy-intensive that in order to heat a hundred-room hotel, you would need to draw energy equivalent to five hundred hair dryers (about 500 kW) all plugged in and turned on at once. As you can imagine, that causes an enormous load on the grid, and grids are not designed for such a large spike in load. So, the only other way to do it is to go back to older methods where you store a large quantity of water and keep it heated—which can be inefficient and unhealthy.

Since on-demand electric water heaters for large commercial properties wasn't feasible in 2009, we had to choose to use natural gas or propane as our fuel. However, we designed our units to be so efficient that they actually cut emissions 40–70 percent compared with contemporary devices. When we started the company, California was debating whether to implement very strict restrictions on NOx emissions in four years' time. We decided from day one that our devices would not only put out less emissions to begin with—70 percent less—but they would also conform to these advanced standards, even though those standards were not yet implemented, just being debated. Our products were years ahead of emissions standards.

Our vision had always been to develop devices that truly have net zero emissions, but we had a challenging problem with combustion we needed to solve. I have always admired NASA and their work in this area, so as we were trying to solve this challenge, I reached out to them. I thought, "If anybody has solved this, it might be the guys at NASA."

I came across a patent owned by NASA for measuring the properties of hydrogen. Hydrogen combusts so quickly that it is difficult to establish proper properties. Then, I got to know and became good friends with Viet, an engineer at NASA who was known internally as "Mr. Combustion." He had developed a new type of burner that burned not only hydrogen but also a range of other gases while putting out less emissions. We ended up essentially licensing that patent from NASA exclusively so that we could convert our devices into even cleaner burning devices and eventually convert them to hydrogen and achieve zero emissions. For this to become a reality, the United States needs a robust hydrogen distribution system or cost-effective hydrogen production technology on-site so that hydrogen (or perhaps a blend) becomes widely available.

Today, with readily available solar technology, the time finally seems right to start moving toward the all-electric units we dreamt of building back when we started the company. But going all-electric doesn't solve all the issues. Electricity still has to come from somewhere—and we can't just keep building larger and larger power plants. We have to have some kind of distributed production of energy supported by localized storage and virtual power plants. Every home would need to be outfitted with a smart electric panel.

Producing and consuming electricity locally also minimizes the typical losses sustained in transporting electricity. For example, today, if you produce electricity in a power station, it is most likely produced by burning fossil fuels. In 2022, 60.2 percent of electricity generated at electricity generation facilities in the United States was from fossil fuels—39.8 percent from natural gas, 19.5 percent from coal, and the rest from petroleum and other gases.[8]

For every one hundred units of gas burned at these power plants, you get about seventy units of electric power. Thirty percent goes to waste. By the time that electricity is transmitted to your home, you end up getting maybe thirty or forty units of electric power, because another thirty or so units are lost in transit. It is very inefficient; it is far better to produce and consume electricity locally. It's the same reason you often see farmers encouraging people to buy local: if you consume local groceries, they don't have to be shipped. The farmer saves on transportation, and they don't have to freeze the produce and hold it and preserve it, so there's no electricity cost. Plus, it's healthier for you to eat fresh food, so it is beneficial all around.

Restructuring power generation and distribution is going to become critical in the coming years. We simply cannot build enough power plants and locate them away from our homes. We have to generate power on our roofs, on solar farms, on wind farms, and so on. The only way we can generate sustainable energy is to harvest free energy from renewable sources like the sun, the earth, water, and wind. Those elements will be around for billions of years. Anything else we harvest has a consequence. The issue with these sources is that they do not produce constant power throughout the day. Thus, the energy generated has to be stored for consumption—a small but surmountable problem.

And the fact is, these are problems we need to solve in order for human beings to survive.

Essential Elements

In order for human beings to thrive and sustain, we need clean air and clean water. It is essential to us as living beings.

Our first act after being born is to breathe in, and our last act when we die is to breathe out. The only air we have to breathe is the air available on this planet, so we are all deeply connected to this planet and its atmosphere. Without clean air to breathe, human beings cannot survive.

As much as we need air, we need water. There is not a soul on the planet who doesn't need to drink water, take a shower, or wash their dishes or clothes with water. Every person on the planet uses water. And as every country in the world has developed or is developing, a basic standard of living is such that clean water or some kind of treated water is a basic human need and a basic human right. It is fundamental to our sustenance.

Unfortunately, both water and air are more contaminated now than they have ever been. Not only are most of our drinking water piping systems a hundred years old, we are also pumping pollutants into the ground, air, and water. This is how crises like the one in Flint, Michigan, are created: lead leaching from old pipes into drinking water.

We have seen the effects of poor water quality firsthand with clients around the country. We got a call a couple of years ago from a client in Texas, saying our tankless water heaters were scaling up because of hard water. We looked at the units, and the problem was not scaling; it was some kind of orange and white powdery substance. It turned out, because the pipes in the building were so old, the water was being dosed with a certain chemical, AQUADENE SK-7100 (a granular blend of ortho and polyphosphates), to prevent the lead in the pipes from contaminating the water. However, they were overdosing the water with this chemical, which was causing the buildup in our units, which looked like hard water scale.

There are examples like this throughout the developed world. I am sure many of us can recount the story of Erin Brokovich, who was instrumental in building a case against Pacific Gas and Electric Company involving groundwater contamination in Hinkley, California. In that case, drinking water was contaminated with hexavalent chromium or chromium 6. In some ways, the less developed world might actually have better water, because they tend to have less contamination from industrial pollution, and they are drinking water directly from the source. I certainly did from our own wells in our house in rural India. The risk here is more often microbial contamination than pollution. In the developed world, you are almost never drinking water directly from the source, and it is usually treated for microbes. However, there is a risk of contaminants from pollution, such as heavy metals like chromium 6. Even if it looks clean, it might not be.

Water, Air, and Earth

Thanks to my father's farm, I grew up with a great awareness of the importance of water. Our land was in a part of the country known as the Granary of South India—a small stretch of the country that produced a disproportionate amount of India's grains, which was mainly rice. It is a very fertile area, irrigated by the river Cauvery and its tributaries, forming the Cauvery delta in the state of Tamil Nadu, India.

To grow rice paddies, you need one to two inches of standing water in the fields. On our farm, that water came from the river, and the river was fed by the rain. If we didn't have a good monsoon season, the rivers would run dry; if there was no water in the river, the wells dried up—and so did our land. One was really at the mercy of weather!

The river Cauvery has its origins in the neighboring state of Karnataka, which generally got a good amount of rain. The river Cauvery would be fed from the rains in the flow from the mountains in that neighboring state and bring water to our area. But if Karnataka did not get heavy rains, it affected us deeply. The neighboring state has a dam erected, and if they got enough—or too much—rain, they would open up the dams. If they didn't, they would close the dam in order to conserve water—which meant my state did not get water. There were always politics at play between the two states over whether to leave the dam fully open or keep it closed, when and how much to adjust it. They couldn't keep the dam shut all the time because it could overflow and flood, but they also couldn't keep it open all the time because all the water would run out of their state.

There were also regulations about how you could draw water from Cauvery and its tributaries, regulated by the water passing through what is essentially a sluice gate. One time, my uncle was so desperate for water for his crops, he tried to cut a different channel to the river—and boy did he get in trouble for that!

If there was no monsoon and therefore not enough water in the river, we had to find a way to irrigate our fields to get enough water in our rice paddies—acres and acres of land. To do this, we had borewells, which are made up of three shallow bores, each about six inches in diameter, bored into the ground about thirty or forty feet. These three were manifolded into a single suction line and driven by a large three-phase motor. The output of this pump was sometimes almost a foot in diameter. From this system, you can pump large quantities of fresh water from the ground to irrigate your fields. The longer the river was dry, the lower the water table would fall, meaning the borewells would have to be sunk deeper and deeper.

Of course, even if the water table was high, the pump worked only if it had good, reliable electric power—which can sometimes be difficult to get in India. If you were in a situation where you needed the borewells, you had to hope the electric power was running strong and stable.

Speaking of wells, when we moved homes, there was no municipal water line, so we had to dig a well. We could always tell how high the water table was by looking down into the well and seeing how far down the water was. From that, you could make a correlation very quickly and predict whether there was water flowing in the river, even though the river was over ten miles away from our house. Normally, the water in the well was about fifteen feet below the lip of the well; if it sank to thirty or forty feet, you knew the river was drying up. Looking at the well was a daily event to keep tabs on what was happening with the river locally and more broadly with the neighboring state of Karnataka.

There were many factors that could prevent us from having a good harvest: if the monsoon didn't come timely or strong enough, if the neighboring states didn't get rain, if there were some interstate politics at play, or if there was some "load shedding" events that made electric power unreliable in running the borewells. And no matter what we did, we were always at the mercy of the weather—something we couldn't control. As such, we were very attuned to changes in the weather, and there was always conversation about how important water is. I grew up with a high awareness about the environment thanks to my father's family farm.

As long as I've had an awareness of waste and efficiency, I've also had an awareness of pollution. When I was young, I had a kind of asthma that was triggered by smoke, specifically automobile exhaust and unburned gases. I could usually tell how well a car or motorcycle was tuned when driving behind it and unwittingly smelling their exhaust as they shifted gears. Incorrect timing, missing plugs, bad fuel—all of these show up in the tailpipe. I was like a walking human emissions detector!

Unfortunately, to this day, if I am stuck behind a car or a truck and their fuel system is not tuned, the exhaust with unburned gas will cause me to develop a chest congestion. Because of this, I've always been attuned to the exhaust coming out of cars and other vehicles, and it was clear to me from a young age that those emissions could not be good. "If this is bad for my lungs," I thought, "it probably shouldn't be in anybody's lungs. It probably shouldn't be in the air." It made me much more aware of what we are breathing in.

This awareness of pollution also extended to the earth. I remember having an Earth Day at RSK (my school) early in my childhood, when we would go and clean up garbage in natural areas. I remember thinking, "I really like keeping things clean. This planet gives us so much. It gives us a home, and water, and air. We shouldn't abuse it. We should keep it clean." In fact, if one were to look at our planet from outer space, it is striking how exceedingly thin our atmosphere is over the earth, almost like a delicate veil against the black backdrop of space.

Out in the community where the farm was located, in the south of India, we had a big festival specifically honoring Mother Earth, called Pongal. In my home state of Tamil Nadu, Pongal is celebrated around January 14, to coincide with the harvest season and the end of winter solstice, and it lasts for three days. The first day is dedicated to the sun; the second day is dedicated to the earth; and the third day is dedicated to cows, which have a great significance in Indian culture.

Over the course of this festival, you thank the sun and earth and cows for all the sustenance they give us. As a child, I very quickly picked up how these elements constitute a complete ecosystem. The sun, water, and earth enable our rice paddies in our fields. That rice and the hay feed us and our cows, respectively. The cows produce milk and butter, which also feed us, and the cow dung fertilizes the rice paddies. Nothing is wasted.

All of these experiences in my childhood gave me a very strong connection to the planet and the environment. I grew up understanding that what we did to the atmosphere was of the utmost importance; that we have to protect the earth—not abuse it—and that we have to use water and energy wisely, not waste it.

After all, we all live on this planet. From coast to coast, from Asia to Europe to Africa, we all share this environment, and we are all going to be affected by changes to it. We are all connected. The COVID pandemic made this interconnectedness even clearer. No matter where you were, if you lived in a mansion or a shack, if you were rich or poor, if you were a white-collar or blue-collar worker, it didn't matter—everyone was affected.

When I lived in Peoria, in Midwest Illinois, which has a huge farming community, I heard someone say, "Farming is everybody's business." That really struck a chord with me. If no one farmed or if crops fail, people don't get to eat—whether they consider themselves connected to farming or not. And if the environment is damaged, farms will fail.

We are already at a level of damage to our air and water, beyond which lies a point of no return. Renowned environmentalist David Suzuki illustrates this crisis using a test tube and bacteria. He asks us to imagine that the earth is a test tube filled with nutrients, and human beings are bacteria in that test tube, living off the nutrients. When you start with a very tiny quantity of bacteria, there will be plenty of nutrients in the tube. Let's say that the bacterium doubles every minute. If the test tube were half full at the fifty-ninth minute and you talked to the bacteria about an impending crisis, they are likely to say, "What are you worried about? We have been around for fifty-nine minutes and still have half a test tube full of food." Of course, within a minute, at the sixtieth-minute mark, the bacterium doubles again and fills the entire test tube, consuming all its resources.[9]

We are around that point with the resources on this planet. We have passed the fifty-ninth minute. The population has grown exponentially, and with the way we waste water and energy and pollute the atmosphere, the situation is dire. It is, without exaggeration, the most pressing problem of our time and for humanity.

The Big Impact of Small Actions

It may seem like too enormous a problem to tackle—but big problems can be solved if many people take small actions. This phenomenon is perfectly illustrated by a famous story about former American Airlines president and chairman Robert Crandall. Looking for ways to save the airline money and fuel, Crandall decided to remove one olive from every salad served on American Airlines flights. This tiny action, repeated hundreds of thousands of times, resulted not only in reducing the cost of food but also in reducing the amount of fuel needed for flights due to eliminating the combined weight of all those olives—to the tune of $100,000. Who would have thought a small olive could have such a big impact?

I firmly believe that a large amount of small, collective actions outweighs fewer larger actions. Obama's campaign came together not because of many large donors—although there were a few, of course—but because of a grassroots movement of thousands of people chipping in a few dollars here and there. There are also many examples of this phenomenon in the business world, such as Uber, Airbnb, and Netflix. These are large companies that have a large impact—but their structure is made up of many individual people doing small things that add up to something large. The average Uber driver doesn't drive all day but only a few hours—but together, all the drivers create a system where there is always a ride available, which is what built Uber into a multibillion-dollar company.

This is exactly how change happens: many individual people taking small actions to create big change. This same thinking, this same phenomenon, can be applied to climate change and water conservation. As I talked about back at the beginning of this book, most people have to wait thirty to forty-five seconds, even up to a minute, for hot water to get to their faucets—and all the while, good, clean water is running down the drain. If we added up those tiny amounts of time, it would amount to six months of wasted water over the course of a lifetime.

Now multiply that by a city of 340,000 people, and that's two Olympic-size pools of water wasted. Along with the water being wasted, there is also the energy being used during that time to heat the water and bring that water to our homes. If we eliminate this energy wastage, we can cut another 1.1 billion pounds of CO_2 emissions—not to mention, eliminating 2,600 Olympic-size swimming pools of good, clean, treated water going down the drain every day.

The small action of not wasting water will help the environment, which creates a healthier atmosphere, which preserves weather and rainfall patterns, which allows farmers to continue farming and providing food to the world. If we don't take these small collective steps, then we may not have enough to eat, and our planet may become uninhabitable.

There is no one exact scientific path to a sustainable future—but there are many of these small, obvious actions we can take. We can switch to electric cars and more efficient lightbulbs. We can recycle and use reusable bags. Those may seem like small, insignificant actions, but when you add them up collectively, the impact is larger than the sum of its parts. For example, I am sure most of us have accidentally left lights on in our houses that don't need to be on. If we turned off those lights, it would eliminate the same amount of CO_2 emitted by burning three billion pounds of coal. If we all inflated the tires on our cars to the proper pressure, it would cut down fuel consumption by about 2 percent, which directly translates into two million gallons of fuel saved—and, in turn, might even impact the cost of gas by lowering demand.

Food waste is another example of how small actions can add up to big change. A large quantity of the food we buy or cook both at home and at restaurants ends up being thrown away. The US Department of Agriculture estimates that a staggering 30–40 percent of the food supply in the United States is wasted.[10] If we all decided to buy or cook smaller portions of food, we would reduce waste and could cut almost ten billion pounds of CO_2. Moreover, if we waste less food, then we don't have to produce or transport as much food, meaning farm machinery can run less, meaning it consumes less fuel and puts out fewer CO_2 emissions. The small action of deciding to buy or cook smaller portions, multiplied by millions of people, would have a dramatic impact on the world.

Even the tiniest of actions, multiplied by eight billion people, equals enormous action—and that is the only way we are going to get out of this crisis that affects us all, equally.

Change Starts by Meeting People Where They Are

We can all do something, today, that can have an impact on climate change. But changing behavior is hard for an individual, let alone for every person on the planet. We need to figure out how to meet people where they are in order to make real change.

That's why, at Intellihot, we wanted to develop a system that would not require people to change their behavior. We knew, in order to make a big impact, we had to conform to people's habits—which is why our machines learn human patterns. We can provide exactly what people want, without requiring any change in behavior, while consuming less energy. That became one of our internal strands of DNA: doing more with less.

Our smart water heaters can learn your patterns in seven days. When you take showers every day, the controls internally record when you used water and how long you used it for. Then, it deciphers, "Today is Monday. Someone used hot water between six and seven o'clock in the morning, then again between seven and eight o'clock in the evening, then from twelve to one o'clock in the afternoon."

The machine uses a twenty-four-hour clock on a seven-day pattern, so when the next Monday rolls around, the machine knows, "You took a shower last Monday between seven and eight o'clock." Therefore, it is likely that you will use water at the same time again. It will heat water and push it all the way to your faucet at that time, so when you open the faucet, you get instant hot water and don't waste any water waiting for it to get hot. If your pattern was different on Tuesday, it learns that pattern, so it can predict your needs every day of the week.

When I had one of our tankless water heaters installed in my house, I would go down to the unit and watch the pattern to see what it had learned. In our house, it showed a few bars between six and eight o'clock in the morning and then nothing, a few bars between eleven and twelve o'clock in the afternoon and then nothing, and then a few bars between five and nine o'clock in the evening. It was the same pattern Monday through Friday. On Saturday and Sunday, however, there would be nothing until eleven o'clock in the morning since we all had a slow start on the weekends!

Of course, your schedule won't necessarily be exactly the same every week of every month of every year. That's why the unit is always learning and always adjusting. Each day, it records water usage and compares it with that day the previous week. If your pattern changes—if your work schedule shifted or if you're going on vacation—the heater will relearn and adjust to your new schedule. In addition, the unit heats on demand; the rest of the time, it behaves like a regular tankless water heater, meaning if you opened the faucet for hot water at a nonregular time, it would still heat water only when you need it.

The only change people need to make is to install our system—and to be open-minded about a new way of heating water. Intellihot water heating units are being used by thousands of customers, and through that usage alone, we have cut around twelve billion pounds of CO_2 emissions—equivalent to eliminating all the cars in America's third-largest city, and my favorite, Chicago.

We introduced this self-learning technology in our products in 2012, and just last year, I noticed that Rinnai, one of the world's largest companies making tankless heaters in Japan, introduced almost exactly the same self-learning technology. It was gratifying to see someone was finally copying us, ten years later! As they say, imitation is the sincerest form of flattery. The industry is finally starting to realize that we shouldn't waste water, and therefore, machines should be self-learning.

Getting Businesses On Board

While individual action can create big change, it's even more effective in conjunction with innovations in the industrial and business worlds—which means we need businesses and industries to get on board with making small changes.

As counterintuitive as it might seem, when it comes to businesses, focusing only on solving the climate problem will not yield rapid change. While perhaps some businesses are thinking about the climate crisis, I can assure you the majority are not. They are not thinking about a system that can lower CO_2 emissions; they are thinking about a reliable product or a business that will make them money.

In order for change to happen, it has to be fundamentally supported by economics. If it is not supported by economics, no amount of marketing or goodwill is going to make it happen—because at heart, businesses are not charities. Their mandate is to create products, have a strong revenue stream, make a profit, and provide a return for the shareholders. Corporate responsibility, ethics, and social impact—all of those are a part of it too. But at the foundational level, to get started, it has to make terrific economic sense. It is good when solutions have a positive environmental impact and are sustainable, but it is even more effective if they have a meaningful impact on the business's ability to make money.

In my early days as an entrepreneur, I intuitively knew that any solutions to the climate crisis we wanted to offer had to be economically compelling to our customers. On its own, being greener is not a selling point or a competitive advantage—this was especially true when we were starting out. Back in 2006, energy conservation wasn't something people were really thinking about, except in terms of saving money. At that time, most people believed that in order to be more energy-efficient and go green, you had to spend more money, even if you were saving on energy costs. For instance, if you bought a hybrid car like the Prius, you would have to drive it for roughly fifteen years before you made your money back on the gas savings.

With our product, that wasn't the case. Here was a product that saved energy, cost less to own, and cost less to operate. It actually created positive cash flow, right from day one! Sometimes, in order to do something better and faster, it doesn't need to be more expensive. You can engineer and innovate your way into something that is economically compelling while delivering solidly on energy efficiency. Our product cuts down emissions, but it does not cost more; it doesn't have the "green premium," as they call it. Creating innovative sustainable products without a green premium should be the number one goal for every engineer in this field.

The fastest way to start making a change is to solve the economic and business problems first. If you reframe the challenge and make it clear that it doesn't have to cost more—that in fact, it can save money—businesses will get on board. And the truth is, economic problems and climate challenge are just different sides of the same equation. Less emissions equal saving energy, equal lower operating costs, and equal saving money. The result of energy efficiency is saving money *and* not having as many pollutants in the air. When the Hilton hotel in Burlingame installed our system, they cut their gas bill by over 70 percent—and also eliminated several million tons of CO_2 emissions that would have otherwise been emitted into the atmosphere. Marriott Corporation has embraced our tankless water heating system design because they believe in overall sustainability, health, and wellness of their customers—in addition to requiring reliable and cost-effective solutions.

We don't need to waste resources to get the same benefits and enjoyment from them. From a moral and ethical standpoint, waste is not good—and from a monetary standpoint, too, waste is not good. Why would you waste good, clean, treated water? You'll just end up paying for it!

If we tackle the problem of climate change from this angle—with innovations that can satisfy commercial needs, with compelling economic value propositions, and that also help the environment—we will be resolving both problems with one solution. If we can create solutions that meet people where they are—for both businesses and individuals—then all those independent changes, on the personal and commercial levels, will add up to something substantial and impactful. If we do this individually and collectively, we can actually achieve an enormous reduction in energy usage, saving individuals and businesses money while also saving the planet and ensuring our future survival.

CONCLUSION

Climate change isn't someone else's problem; it's *our* problem. We can't, and shouldn't, wait for one great, big, easy-fix answer—because there isn't one. But you, reading this book, no matter who you are, no matter what your position in life—whether you are an entrepreneur, a business owner, a CEO, a sustainability officer, or just a private citizen in your individual life—can do something. And you should, because we live under the same sky, and, whether we like it or not, we are all pallbearers in the climate journey of our planet. Each and every one of us is responsible for our collective fate.

Perhaps you are in a role where you can make a difference by integrating our products and designs into your building—CEOs, business owners, engineers, architects, builders, or operators and people of influence in the lodging, multifamily, senior living, education, health care, or restaurant facilities. If you are, you can immediately make a huge difference by making a simple change to your water heating systems—while maintaining and even improving your bottom line. I believe solutions to climate change must be fundamentally supported by economics, and the only way to really make change is to solve the economic and business problem first. Intellihot does both.

If you want to discuss how to get started switching your water heating systems over to Intellihot, simply contact us at 1-877-835-1705.

Our water heaters have already cut twelve billion pounds of CO_2 emissions, equivalent to eliminating all the cars in Chicago, and we have barely scratched the surface. Imagine the impact if more businesses adopted our technology or something similar.

Perhaps you are an entrepreneur, following your own purpose and calling. To you I say: we are the crazy ones, crazy enough to believe we can change the world—and we often do. So, keep going, be resilient, find that next ounce of strength, solve that problem—and yes, it is worth it. You will create jobs, change lives, create immense wealth, and make the world a better place. What has worked for me, and what I believe, is that success, at its core, comes down to three things:

1. **Find something worthwhile**—you will know it when you do.

2. **Believe that it can be done**, no matter how big or seemingly impossible it is.

3. **Stay hungry, stay grounded, and keep rolling.**

There is a famous quote attributed (somewhat dubiously) to Henry Ford: "If I had asked people what they wanted, they would have said faster horses." Nobody could conceive of an automobile—except for an inventor and entrepreneur who had that idea, went out, made it happen, and changed the world forever.

If you have an idea, go out and make it happen. As an entrepreneur—or an entrepreneurial-minded person working at a big company—you are someone who can map the future. If you have an amazing idea (or ideas!) about how to change the world, send me an email at sri@deivasigamani.com and let's talk about it.

To each and every person reading this book, I want to say: You can change the world just by doing something small. By making small changes, an effective and meaningful change is possible because, most certainly, a large collective of small actions always outweighs fewer large actions.

Let's stop climate change—starting right now.

The Intellihot Water Heater

The water heaters we have developed at Intellihot are truly distinct from anything else on the market. Current water heaters come in two basic designs: (1) tank and (2) boiler with a side-arm tank. Neither design heats water when needed (on demand) nor is effective at serving the market in terms of efficiency, reliability, cost, and space. First, these systems store large volumes of hot water 24/7 even when water is not being used, resulting in large standby losses over time. Second, they have poor heat transfer because of their system design, and frequently, the efficiency of heat transfer is further reduced from a buildup of sediment and scale.

Intellihot's water heating systems are designed to overcome these deficiencies and realize the full potential of on-demand technology while delivering unparalleled reliability and savings in energy and space.

What Should an Ideal Hot Water System Look Like?

An ideal water heating system should be just like your car: it should be started when you are ready to go, not left idling overnight, and it should be reliable. It should start up in rain, sun, snow, or any inclement weather. And it should run on fuel of any octane rating.

Unfortunately, current water heating systems are akin to using a bulldozer to commute to work: big, bulky, and left running. To make matters worse, sometimes two bulldozers are installed for the sake of redundancy and reliability.

The following are some desired characteristics of an ideal water heating system.

It should handle varying loads with precise temperature control. The human body can detect changes in temperature within ±4°F, and this was one of the reasons why tanks are used. A water heater should be robust to thermal cycles and poor water quality. These two parameters are par for the course in any application. If one were to look at the map of the United States, it is apparent that most of the water is classified as hard or very hard. The dissolved solids from water can accumulate and cause poor heat transfer and result in lower efficiency and susceptibility to failure. The gas train and systems should be able to function well under low gas pressures.

In places like New York, especially during winter, a surge in demand or switching on another appliance could cause a dip in gas pressure. Failure to ignite is usually the number-one annoyance for building maintenance engineers. The heater should be easy to install and operate. Being compact is very helpful as it can be maneuvered into tight spaces or taken up through standard elevators and doorways. Cutting roofs open and using helicopters to drop in equipment can be expensive. Grid connectivity should be standard. The grid of the future will most certainly need distributed energy storage and demand response. Finally, self-learning systems—ones that adapt to the behavior of the user—would be ideal. This way, we can maximize efficiency without sacrificing comfort. In addition, wastage of water can be avoided.

Why Are Storage Tanks Used?

Storage tanks have been used since time immemorial. Metal hot water storage tanks have been around since the Industrial Age. There are two main reasons for storage tanks: First, there is significant uncertainty in how much hot water is needed and when it might be needed. There have not been any large-scale measurements of hot water consumption in commercial buildings like hotels, multifamily dwellings, hospitals, and the like. This uncertainty on how much water is needed leads to storing more water than needed so that it is "just" there for use. Second, storage tanks are deployed to compensate for the lack of performance by traditional systems—inadequate temperature control, inability to handle cycles, thermal shock, or turndown.

Both of these reasons lead to an enormous storage of hot water and also a doubling up of equipment. In the United States today, nearly six billion gallons of water are kept hot 24/7, roughly one-sixth of Lake Tahoe. Twenty-eight percent of the world's energy is consumed by the building, making them the largest consumers of energy globally. Twenty-five percent is used for water heating. Intellihot cuts this energy use by 40 percent.

Autonomous Masterless Controls

Currently, water heaters and boilers installed in a cascading situation are paired using a master-slave approach. Essentially, in this case, a single master controller is responsible for scheduling all the units. However, this creates a single-point failure: What happens if the master controls fail or if a rat chews up the communication cable coming from it? Everything today in the digital world, from computer servers to industrial controllers, is predominantly set up as master-slave control.

Intellihot uses a propriety patented method called "masterless cascading." This system eliminates the need for a master controller and single-point failures. In addition, the Intellihot units share information on operational hours, firing cycles, flow rates, and more. Then, they use an algorithm to distinguish city driving versus highway driving and automatically rotate units to normalize wear and tear. Rotation by hours is simply inadequate. This invention required Intellihot to invent new protocols of machine-to-machine communication and handle collisions of data. Today, all of the systems utilize this method.

What Is telliSize?

Intellihot developed a sizing tool called telliSize. It is based on real-world field data, and after evaluation of ASPE and prevailing sizing methodologies, Intellihot developed this sizing calculator to "rightsize" the capacity needed per application. This capacity is then broken into multiple units to result in a system that has excellent performance under varying loads while simultaneously leapfrogging reliability and energy savings. When rightsized, storage can be eliminated entirely.

Electric heat pumps are quite challenging to size and implement. This is because their performance is tied to numerous time-variant parameters, such as outdoor air temperature, seasonality of weather, groundwater temperature, and finally the varying nature of water consumption itself. It is most definitely not a one-dimensional problem. Failure to properly account for these variables will result in only two outcomes: systems being designed too large or too small. What we need is just right, and in order to achieve that, we built a sophisticated simulation tool that takes in account all these time-variant factors to simulate operations over 365 days. The result is a high-confidence, accurate, and guaranteed sizing that maximizes efficiency at the most affordable cost.

Useful Industry Jargon

When it comes to water heating in the commercial and industrial sectors, it is useful to understand some common terminology and how they affect system performance and design. They include turndown, thermal shock, temperature control logic, failure to ignite, and flue as recirculation.

TURNDOWN

Since a majority of hot water applications have varying demands throughout the day, it is implicit that the hot water equipment operate over a wider firing range. If this range is restricted, as is the case with typical boilers (5:1), a storage tank has to be incorporated to overcome this deficiency. Both storage tank and short cycling have a dramatic effect (estimated to be 40 percent) on gas consumption.

The following graph illustrates how a typical boiler and Intellihot operate when the demand is 30,000 BTU/hr.

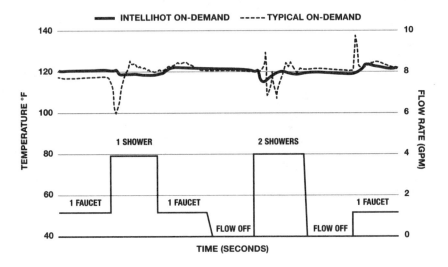

THERMAL SHOCK

Certain applications have a significant number of on/off operations inherent in their business. Full-service restaurants, in particular, are more prone to cycles (one million cycles in nine months) than hotels or multifamily dwellings. Furthermore, if the intended solution for any application is to be "on demand," then that in itself introduces cycles. Hence, any heat exchanger must be fundamentally designed to handle these cycles. Intellihot uses a single-finned tube that is fully floating, thus allowing for expansion/contraction as well for extreme thermal shocks that occur when directly feeding in cold water (no storage employed).

TEMPERATURE CONTROL LOGIC

Human skin is an excellent detector of variation of temperature, typically 5–7°F. Therefore, for "on-demand" hot water, it is essential to maintain temperature variations within this band for anticipated fluctuations in flow. For an analogy about demand variations, see the graph under the "Turndown" section. It represents an application with two showers and two faucets and a typical cycle of turning on/off.

Intellihot uses a combination of temperature sensors and measured flow rate, along with dynamic mathematical model-based controls, to achieve precise temperature control, rapid response, and stability. Classical control methods such as feed-forward or feedback controls are inadequate for a true on-demand system. However, one way to overcome this shortcoming is to build storage around such heat exchanger designs (usually fire tube) to compensate for poor controls.

FAILURE TO IGNITE

Given the growing demand for natural gas, along with seasonal demands and inadequate distribution infrastructure, it is common to see depressed gas pressure during operation in the northeastern United States. Also, other devices connected to the same gas line may cause a temporary or sustained low gas pressure. Failure to ignite and resulting lockout can disrupt business operations.

Intellihot utilizes an extremely powerful premix blower. Every 250,000 BTU/hr of firing rate is supported by an individual blower. In addition, the specially designed swirl plate and gas injector make the gas train uniquely robust to low gas pressures. The unit is capable of achieving full firing rate for gas pressures as low at 2.5" WC.

FLUE GAS RECIRCULATION

Conventional boilers may require the air intake and the exhaust to be at the same planar location so as to provide minimal pressure variations or disturbances from wind, draft, and other natural elements. However, the colocation of intake and exhaust may cause the flue gas to recirculate, damaging the boiler.

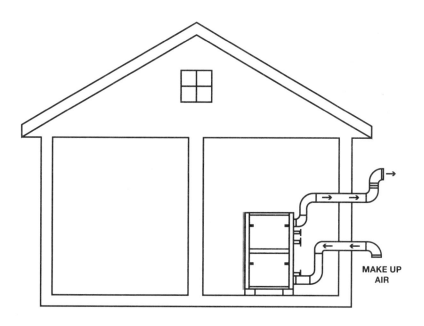

Intellihot allows and encourages the separation of intake and exhaust to prevent this issue.

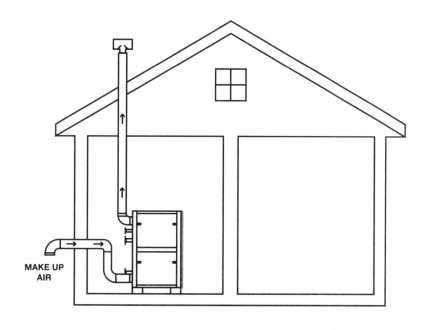

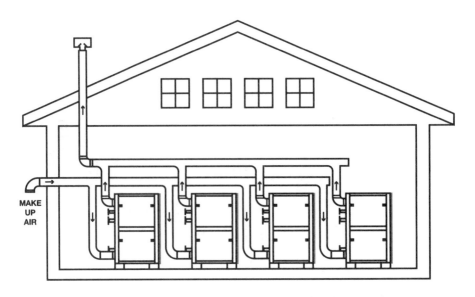

MAKE
UP
AIR

Want to know more about Intellihot's unique engineering?
Scan the QR code below to learn more!

CONTACT

If you want to discuss how to get started switching your water heating systems over to Intellihot, simply contact us at 1-877-835-1705.

ENDNOTES

1   Ethan Siegel, "How many atoms do we have in common with one another?," Forbes, April 30, 2020, https://www.forbes.com/sites/startswithabang/2020/04/30/how-many-atoms-do-we-have-in-common-with-one-another/.

2   Avinash Kumar and John Bechhoefer, "Exponentially Faster Cooling in a Colloidal System," Nature 584, no. 7819 (August 2020): 64–68, https://doi.org/10.1038/s41586-020-2560-x.

3   Smarter House, "Replacing your water heater," accessed May 15, 2022, https://smarterhouse.org/water-heating/replacing-your-water-heater.

4   Caitlin R. Proctor, William J. Rhoads, Tim Keane, Maryam Salehi, Kerry Hamilton, Kelsey J. Pieper, David M. Cwiertny, Michele Prévost, and Andrew J. Whelton. "Considerations for Large Building Water Quality after Extended Stagnation," AWWA Water Science 2, no. 4 (June 16, 2020): e1186, https://doi.org/10.1002/aws2.1186.

5   Nora Knox, "What is WELL?" U.S. Green Building Council, April 2, 2015, https://www.usgbc.org/articles/what-well.

6   M. González-Torres, L. Pérez-Lombard, Juan F. Coronel, Ismael R. Maestre, and Da Yan. "A Review on Buildings Energy Information: Trends, End-Uses, Fuels and Drivers," Energy Reports 8 (November 1, 2022): 626–637.

7   Steve St. Angelo, "How much human labor equals one gallon of gasoline?" TalkMarkets, January 26, 2020, https://talkmarkets.com/content/how-much-human-labor-equals-one-gallon-of-gasoline?post=248539.

8   U.S. Energy Information Administration (EIA), "What is U.S. electricity generation by energy source?," March 2, 2023, https://www.eia.gov/tools/faqs/faq.php.

9   MIT—Docubase, "The test tube with David Suzuki," 2010, https://docubase.mit.edu/project/the-test-tube-with-david-suzuki/.

10 US Department of Agriculture, "Food waste FAQs," accessed February 5, 2023, https://www.usda.gov/foodwaste/faqs.